Valuation of Internet and Technology Stocks

To my wife, Nadia, without whose support this book would not have been written, and to my sister Pat, without whom it would not have been typed

Valuation of Internet and Technology Stocks

Implications for Investment Analysis

Brian Kettell

BUTTERWORTH
HEINEMANN

OXFORD AMSTERDAM BOSTON LONDON NEW YORK PARIS SAN DIEGO
SAN FRANCISCO SINGAPORE SYDNEY TOKYO

Butterworth-Heinemann
An imprint of Elsevier Science
Linacre House, Jordan Hill, Oxford OX2 8DP
225 Wildwood Avenue, Woburn, MA 01801-2041

First published 2002

British Library Cataloguing in Publication Data
Kettell, Brian
Valuation of Internet and technology stocks: techniques and investment analysis
1. Corporations – Valuation 2. Internet – Economic aspects
3. Technology – Economic aspects
I. Title
332.6'3221

Library of Congress Cataloguing in Publication Data
A catalogue record for this book is available from the Library of Congress

ISBN 0 7506 5383 3

Composition by Genesis Typesetting, Laser Quay, Rochester, Kent
Printed and bound in Great Britain by Biddles Ltd, *www.biddles.co.uk*

Contents

Preface

The collapse of Internet and technology stocks

It was a bubble, and it did burst. About that most people agree. The history books will record 10 March 2000 as one of the most tumultuous days ever to engulf the stock market. In a matter of hours billions of dollars were wiped from stock markets all over the world, as the technology bubble finally burst.

America's Nasdaq index, made up largely of technology firms, collapsed into what many believe to be the biggest stock market crash of all time. The Nasdaq Stock Market smoothly rose six-fold to a record high over the five years ending 10 March 2000, before collapsing. This was not the first time stock prices have been driven by waves of 'irrational exuberance' rising to dizzy heights and falling back accordingly.

'Stock prices have reached what looks like a permanently high plateau,' declared the eminent Yale economist Irving Fisher in the autumn of 1929. A few weeks after this oracular pronouncement, the Dow Jones Industrial Average had declined by more than a third. The worst was still to come. On 8 July 1932, the Dow Jones closed at 41.88, a drop of nearly 90% from its 1929 peak. The stock market chart of these years resembles a precipice rather than a plateau.

Why did Professor Fisher get things so wrong? The answer is that he had fallen for the decade's most alluring idea, a thesis which underpinned the great bull market of the late 1920s: he believed that America had entered a new era of limitless prosperity.

Investors trying to make sense of the dot.com debacle might do well to recall the story of William C. Durant. Early in the twentieth century, Mr Durant founded General Motors Corporation, one of a revolutionary group of automakers that in some ways were the dot.com companies of their era. Investment capital poured in and GM stock grew by leaps and bounds from 1914 to 1920. But when the overcrowded automobile industry failed to deliver on inflated expectations in the early 1920s, auto stocks plunged. GM lost two-thirds of its stock value in six months. A panicked Mr Durant borrowed money and feverishly bought the shares in a futile attempt to prop them up.

GM, of course, eventually recovered and soared again – too late for its founder, who lost his entire fortune and wound up running a bowling alley. Many other once-great auto companies, such as Packard, Studebaker and Hudson, fared less well than GM, disappearing one by one in the ensuing decades.

The current technology stocks boom and bust, it turns out, is nothing new. It is simply the latest in a long history of investment bubbles that have plagued shareholders since investing began. From the Dutch tulip bulb craze of the seventeenth century, to the locomotive revolution of the nineteenth century, to the rise, fall and resurrection of personal computer stocks and biotechnology stocks in the 1980s and 1990s, investors have fallen madly in love with – and then madly out of love with – the hot technology of the moment.

History does not suggest that, eventually, many of the pieces of a burst stock bubble get patched back together. Some of the stocks will soar again. But the recovery typically takes longer than investors hope – usually years, not months. Many of the stocks caught up in the bubble will fail to survive, and of those that do survive, all but the most robust will wind up behaving less exuberantly than they did before.

This text does not claim to provide any message with respect to when or if the year 2000 technology stock market collapse marks the beginning or the end of any new phase of economic/ stock market activity. However, the whole momentum of this latest boom–bust phenomenon has opened up many intriguing technology-related conceptual questions for economists, accountants and government policy makers, to highlight just three groups in society. These are questions to which there are no clear answers. In this text I have tried to take readers on a voyage of discovery through the intricacies of many of these issues. Enjoy the trip!

1

The age of information and the democratization of data

> Is it possible that there is something fundamentally new about this current period that would warrant complacency? Yes, it is possible . . . but regrettably history is strewn with visions of such 'new eras' that, in the end, have proven to be a mirage. In short, history counsels caution.
>
> (Alan Greenspan, February 1997)

This book is about the valuation of technology stocks whose ultimate value is linked to the growth of the Internet and the World Wide Web. It is of course written at a time when many of the so-called dot.coms have collapsed in price. It nevertheless remains the case that the ultimate value of many of these stocks revolves around the valuation of intangible assets, assets that by definition are not easy to value. However, I do not share the investment viewpoint that 'if you can't value them . . . you shouldn't buy them', and this book sets out an approach to the valuation of technology stocks. It's a long road, travelling through economic history, rapid technological change, speculative stock market bubbles and the bursting of the same, changes in accounting methodology and the mass psychology of investors, to choose just some of the chapter contents.

To put recent trends into a historical framework, it is useful first to look back in time.

Technological change and the stock market in history

Every major advance in communication and transportation has been a marvel to users, a cause for a huge stock market celebration and, for the economy in most cases, significantly lowered costs of distribution. Ultimately, every distinctive innovation matured and was eclipsed by the next one.

In the late 1600s, Turnpike Trusts in the United Kingdom became very popular with investors and provided dramatically improved efficiencies. As existing roads were narrow and rough, packhorses were widely used. The new toll roads featured a well-structured roadway wide enough that wagons could pass. A packhorse could carry about 250 pounds, but the turnpike could support a two-ton wagon load pulled by a two-horse team resulting in a quadrupling in productivity.

In the 1720s, the prolific writer Daniel Defoe noted that the reduction in freight rates mainly benefited tradesmen rather than carriers. Obviously, conditions were competitive. The public heard all the touts, but considered the tolls as taxation and were impatient for benefits. The following doggerel sums up the party aspects of the boom and its consequence:

> Now with Turnpikes are grown much in fashion the
> hardest Tax in all our Nation – for where Wine &
> Women & Stock-Jobbing past, The Turnpike must
> help us at last.

The jump in efficiencies with canals was outstanding as one horse could tow a barge carrying 50 tons. The improvement over the packhorse was by a factor of 400 times or over the turnpike by about 50 times. Speculators frantically chased new canal issues.

Of course, the construction time from stock mania to operating success was measured in years and canals provided the first great vision phenomenon well in advance of commercial reality. A new issue mania erupted in 1792 and subscriptions could only be entered with the company in the town where the canal was to start. Newspapers that winter provided entertaining accounts of speculators wildly galloping through the night in snowstorms to get to the next town for the start of business.

As wondrous as they were, canals had limitations, with droughts in summer and ice in winter. Also, speed was limited due to bank erosion from the wake of faster 'packets' designed for passengers and mail.

Horse-drawn freight wagons rolling on wooden rails had long shown efficiencies superior even to wide (up to 9 inches) steel-tired wheels on improved roads. Originally developed as feeders to turnpikes and then to canals, all that was needed was a mobile steam engine to launch another revolution in communications.

Naturally, speculators were in full song well before commercial success. The high-tech concept celebrated in the bubble that peaked in 1825 was the railroad. 'Nothing now is heard of but railroads,' said the *Quarterly*, a London publication, in May 1825. 'The daily papers teem with notices of new lines in every direction; and pamphlets are thrown before the public eye recommending nothing short of them general throughout the Kingdom.'

During 1824 and 1825 prospectuses were issued for 624 companies, and in March of 1825 the famous merchant banker, Francis Baring, noted 'The gambling mania [has] seized upon all classes and was spreading in all parts of the country.' That bubble topped out in the summer and the stock market was into severe speculative frenzy in late October when the first commercial railroad began operation. Railroads did not reach ultimate saturation in the United States until the 1890s, just as the automobile was being developed. Automobile sales collapsed in the severe recession following the1920 commodity boom. They spiked up again in the stock market euphoria just prior to the 1929 stock market crash, when they again collapsed.

The electronic world, so to speak, began in the 1860s with telegraph using Morse code and by 1869 undersea cables provided instantaneous communication between America, Europe and England. All of this was celebrated as high-tech in the 1873 Bubble. Obviously, the telephone, in moving from Morse to voice, was an outstanding but evolutionary step. The effect of the introduction of the telephone as a distinctive jump in communications and transportation was highly celebrated in the 1920s' great financial boom.

Radio communication using Morse was expanded by military purposes during the First World War and handed over to RCA. The business plan in 1920 could not have imagined network broadcasting of voice and music, as well as gramophones and talking pictures by 1929. Saturation in both the stock market and the industry was reached in 1929, and the subsequent market crash is well documented.

The defining moment for the Internet: 'The Protocol for Packet Network Intercommunication'

Until the 1970s the idea of the world as a single, unified entity was no more than that – an idea. It began to become a reality with the development of computer technology and information theory that was to lead to the creation of the Internet. This was a process of immense technical complexity over which there was no one presiding genius. One of the Internet's strengths is that it is not dependent on a single insight or technology. But it is fair to say that the May 1974 edition of the obscure technical journal *IEEE Transactions on Communications* was a defining moment in twentieth-century history and, probably, the new millennium also.

The paper published by Robert E. Kahn and Vinton G. Cerf was unpromisingly entitled 'A Protocol for Packet Network Intercommunication'. It is technical, jargon-laden and as dry as dust. It concludes: 'In particular, we have described a simple but powerful and flexible protocol which provides for variation in individual network packet sizes, transmission failures, sequencing, flow control, and the creation and destruction of process-to-process associations.' In short, they had defined the Internet. The key to the Net is that it is, indeed, a net. It has no centre; rather it is a connection system between millions of computers. As a result, it is impossible to turn off unless there is a total global failure of electricity supplies and, in addition, it is formidably resilient. Thanks to the way 'packets' of information will always find their own routes through the Net, it can survive the destruction of much, or even most, of its fabric.

The power of the network is that it lies beyond the control of its users, companies or nations. Like roads, railways and telephones before it, the Net brings a hitherto inconceivable level of connectivity to the world. But unlike them, it also creates a new world – cyberspace. This is a 'place' in that it can be navigated, and the language associated with it is that of geography or architecture. So there are chat 'rooms', web 'sites', Net 'locations' and we are said to 'surf the Net' as if it were an ocean.

The impact that the Cerf and Kahn paper was to have was not fully recognized at the time. However, the authors were well aware that, with hindsight, some past forecasts about technology and its economic impact had been wildly optimistic. Cerf and Navasky, in a later paper, quote Thomas Edison's 1910 forecast that 'the nickel–iron battery will put gasoline buggies out of existence in no time' (1984, p. 229). But other forecasts were equally wildly pessimistic. Cerf and Navasky also quote the president of the Michigan Savings Bank's prediction in 1903 that 'the horse is here to stay, but the automobile is only a novelty, a fad' (1984, p. 228).

The democratization of data

A group of researchers at the University of California, Berkeley, led by Peter Lyman and Hal Varian, collated the estimated output of unique information the world is currently producing each year. The figure they came up with, in 1999, is about 2 exabytes. (An exabyte is roughly a billion times a billion bytes, or the equivalent of around 20 billion copies of a business magazine.) The authors of the study point out that soon it will be technologically possible for an average person to obtain access to virtually all recorded information.

To calculate the numbers the authors start with examining the four main storage media: paper, film, optical disks and magnetic devices such as disks and tapes. Take film, for instance. More than 80 billion photographs are taken each year around the world, according to America's Department of Commerce. UNESCO puts the number of movies produced annually at 4250. And then there are 2 billion x-rays.

The researchers translated these numbers into bytes – 5 megabytes per photo, 4000 megabytes per movie, 8 megabytes per x-ray, for example – and added them up. To do this, they made certain assumptions about the degree to which digital information is compressed to save space.

The main conclusion from analysing this data is what the authors call the 'democratization of data'. It is individuals around the globe who create and store most of this data – about 740 000 terabytes (thousand billion bytes) a year. Published information adds up to a mere 285 terabytes.

The researchers also calculated the amount of information generated by communications such as e-mail, radio and telephone calls, since all of this will probably be archived systematically in the near future. The results are equally staggering. About 610 billion e-mails are sent each year in America alone, adding up to more than 11 terabytes. But this is nothing compared with the mountain of data generated by telephone calls: 576 000 terabytes.

Kenneth Arrow, in his 1973 Nobel Prize for Economics acceptance speech, spoke of information as 'an economically interesting category of goods which have not hitherto been afforded much attention'. The world has caught up with Arrow's comments. The age of information is truly with us.

2

The New Economy: where do Internet and technology stocks fit in?

> When you see reference to a new paradigm, you should always, under all circumstances, take cover . . . There was never a paradigm so new and so wonderful as the one that covered John Law and the South Sea Bubble . . . until the day of disaster.
>
> (J.K. Galbraith, *The Great Crash*, 1955)

During the 1990s, strong economic growth in the United States, combined with low inflation and a pickup in labour productivity growth, led many people to label the phenomenon a 'new economy'. But there was really little consensus on what was different about the US economy nor on what the term meant. This chapter examines what was meant by the New Economy, providing the background for understanding the complex issues involved in the valuation of Internet and technology stocks.

The New Economy defined

Definitions of the New Economy are not precise but typically include one or more of the following characteristics.

A higher rate of productivity growth

The higher rate of productivity growth is largely related to investment in information technology (IT). This has resulted in what economists refer to as 'capital deepening', the process whereby the quantity of capital per worker increases over time (the capital/labour ratio). Recent examples of this are the investment in information technologies, reflecting falling computer prices and new ways computers can help accomplish old tasks with fewer inputs. The microprocessor lies at the heart of these technological developments. The microprocessor's ability to manipulate, store and move vast amounts of information has, it is argued, shifted the US economy's centre of gravity, creating the era of smaller, faster, smarter, better, cheaper. To gain an appreciation of the implications of these technological developments it is instructive to look back in US history (see McTeer, 1999).

From 1895 to 1915, a great burst of inventiveness ushered in an era of rapid technological change and economic growth. Americans saw the arrival of one marvel after another – automobiles, aeroplanes, telephones, phonographs, radios, elevators, refrigeration and much more. These new inventions barely registered as a blip in a GDP dominated by farming, shopkeeping and small-scale production. In time, though, the industries that grew out of them formed the economic backbone of the twentieth century. The advances of this long-ago era would have been impossible without a technology that arrived just after the US Civil War: electricity. Thomas Edison created the lightbulb in 1879 for the simple task of illuminating a room. To build a market for his invention, Edison harnessed electricity, building the world's first generating plant and a distribution network in New York City. As it spread through the economy, electricity recast the economic paradigm. Without electricity, there would be no spark for internal combustion engines, no power for telephones, radios, refrigerators and air conditioners. Electricity provided an ever-ready energy source for factories, with mass production driving down the cost of making just about everything.

Like electricity, the microprocessor is an important invention in its own right and one that shook the world as it touched off a rapid-fire proliferation of spillovers. In 1958 Jack Kilby of Texas Instruments fashioned the first integrated circuit, a bundle of transistors on a piece of silicon. Thus began the grand theme of modern electronics – ever smaller, ever more powerful. Thirteen years later, Ted Hoff of Intel developed the silicon-etching process that produced the first true microprocessors. Initial applications centred on number crunching and rapid data entry. Handheld calculators arrived in 1972, bar code scanners in 1974 and the personal computer in 1975.

Over the next decade or so, American industry applied microprocessors to other tasks. Whole new products, progeny of the digital electronic revolution, burst onto the market place – cellular telephones, robotic factory hands, air traffic control systems, global positioning satellites, laser surgery tools, camcorders, palm-size personal organizers, to name just a few.

Microprocessors made existing products better, cheaper and more efficient. Starting in the early 1980s, 'smart' features helped fine-tune televisions, cut energy use by refrigerators, control cooking in microwave ovens, memorize programme schedules in VCRs and generate diagnostic reports for automobiles.

Universities were the first to hook computers into networks, but it wasn't long before everyday Americans began to connect via electronic mail. The Internet entered the 1990s as an obscure communications network for educators and scientists. It ended the decade as the library, shopping mall and playground of the masses. The Internet is creating spillovers making existing industries more efficient and spawning entirely new ones, including web page design and Internet service.

Computer processing power leapt 7000-fold in three decades. Number crunching tasks that took a week in the early 1970s now require but a minute. The Pentium chip, released by Intel in May 1993, crowds 3.1 million flawless transistors on a square of silicon 16 millimetres by 17 millimetres. It can churn out calculations at up to 112 million instructions per second (mips).

Data storage capacity and transmission speeds surged right along with the more powerful microprocessors. A single memory chip now holds 250 000 times as much data as one from the early 1970s – the difference between one page of text and 1600 books. Transmission speeds increased by a factor of nearly 200 000. Sending the 32-volume *Encyclopaedia Britannica* on the Internet from New York to San Francisco would have taken 97 minutes in 1970. Today's trunk lines can move the equivalent of eight full sets in just one second.

Great leaps of power, capacity and speed led to even greater reductions in the cost of managing information. Intel's vintage (1970) chips sold for $7600 per megahertz. Today's Pentium III chip supplies its computing power for 17 cents per megahertz. The cost of storing one megabit of information – enough for a 320-page book – fell from $5257 in 1975 to 17 cents in 1999. Sending the *Encyclopaedia Britannica* coast to coast would have cost $187 in 1970, largely because of slow data-transmission speeds and the expense of a long-distance telephone call. Today, the entire Library of Congress could move across the nation on fibre-optic networks for just $40.

As the new technology became better and cheaper, American businesses and households embraced it. Only a few thousand homes had a PC in 1980. Now more than half of US families own computers, the newest of them 200 times more powerful than IBM's first PC, introduced in 1981. Three-fifths of US households are connected to the Internet, a mode of instant communication scarcely heard of at the start of the 1990s. This growth in technology provides the underpinning to the idea that the old rules of economics have been dramatically changed.

A rise in total factor productivity growth

Total factor productivity growth refers to the growth in output that is not explained by the physical increase in either capital or labour. Its contribution to economic growth is not easy to identify and measure but nevertheless is an important source of economic growth. This phenomenon is due to the increased utilization of information technology across the economy with resulting spillover effects. Total factor productivity growth depends on:

- technological change
- other advances in knowledge, e.g. just in time manufacturing
- economies of scale.

Spillover effects occur when returns to an investment increase because others make similar investments. Examples here would be networking and the returns to an Internet-capable computer as more consumers and businesses connect to the Internet.

The effect of the Internet has been to intensify product market competition with associated efficiency.

The potential impact of the Internet gains can be seen from Figure 2.1. It took 36 years to achieve 50 million users for radio, 13 years for TV, 16 for PCs, but for the Internet it has taken fewer than 5.

Technology spillovers: increasing returns and decreasing costs

Even when individual industries face decreasing returns to scale, the economy as a whole may enjoy increasing returns when technology spillovers from one industry benefit others. Technology spillovers are especially abundant with inventions

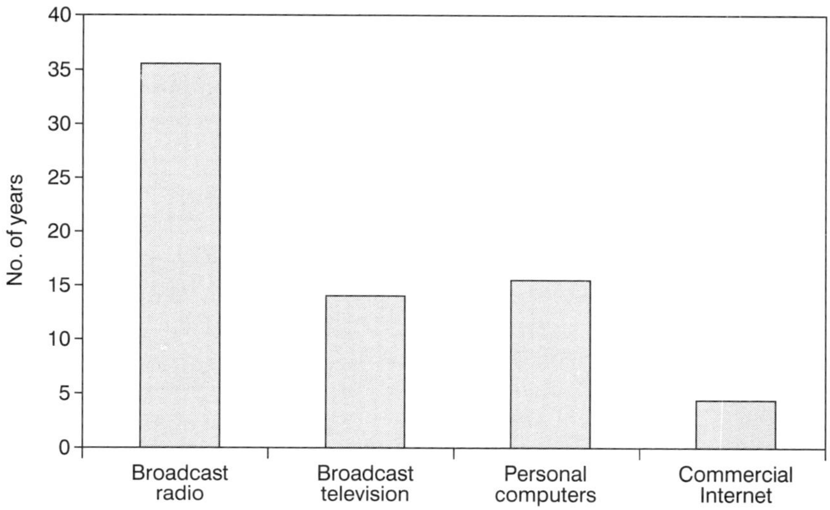

Figure 2.1 Years required by different technologies to achieve 50 million users. *Source:* US Commerce Department

whose applications spread far and wide. Innovation in one company – though intended solely for internal benefit – can spark innovation in others, triggering a powerful, economy-wide cascading effect. Revolutionary technologies can take decades to spawn all their spillovers, during which, for all practical purposes, aggregate returns to scale increase. Examples of this would include:

- Texas Instruments was trying to reduce the size of electronic circuitry when engineer Jack Kilby developed the integrated circuit in 1958. The benefits of that innovation far exceeded what Texas Instruments could internalize, opening a whole new science in which electronic circuitry would shrink to sizes once thought unachievable.
- Intel was pursuing circuitry small enough for a pocket calculator when Ted Hoff developed the silicon-etching process that ultimately led to the microprocessor. A 1971 ad in *Electronic News* heralded the 'computer on a chip' and signalled the start of the digital age.
- In seeking to make microprocessors ever smaller, IBM developed the scanning tunnelling microscope. The benefits of that research, however, went far beyond what was envisioned. The microscope enabled an entirely new industry – nano-technology – that promises to deliver molecularly engineered materials that will reshape our world.

The idea that these achievements have really transformed our lives does, however, have its critics. Gordon (1998b), a leading New Economy sceptic, argues that computers pale in comparison to earlier technological advances like electricity, the internal combustion engine, or biotechnology. Gordon argues that computers may not be exceptionally productive since they primarily redistribute output, not create it. By this Gordon means that computers may increase the utility of workers by providing better working conditions, which would include computer games, or they create output that is unvalued by customers, fancy fonts being one example.

An increase in factor utilization

This is seen most clearly in the decline in the concept of the Non-Accelerating Inflation Rate of Unemployment (NAIRU), a principle first defined by Modigliani and Papademos (1975). Many economists (for example, Meyer, 1997) subscribe to the view that there is some threshold level of the unemployment rate at which supply and demand are balanced in the labour market (and perhaps in the product market as well). This balance yields a constant inflation rate. So NAIRU is that rate of unemployment which can be sustained without a change in the inflation rate. If the unemployment rate falls below this threshold level (NAIRU), inflation tends to rise progressively over time. The US unemployment rate in the late 1990s was widely believed to be below this threshold; hence the puzzlement at the low inflation rate.

A possible explanation of this failure of inflation to rise in the face of strong GDP growth and low unemployment is that the NAIRU had declined; that is, the level of the unemployment rate at which the supply of and demand for labour are in balance may be lower than it used to be. The argument, expressed, for example, by Fed Chairman Alan Greenspan (1997), that technological change has added to workers' insecurity in recent years and made them less willing to push for higher wages, may be thought of as one version of this explanation. Greater insecurity might reduce the upward pressure on wage rates at any unemployment rate and so lower the threshold rate at which wages (and prices) would begin to move upward. The reason why intensification of product market competition should lower the level of NAIRU could be due to the effect of the Internet and to the globalization of world markets (discussed below).

Wadhwani (2000) has demonstrated that the effect of the Internet in the retail market should be to lower prices because:

- lower search costs should lead to lower prices
- lower market entry costs will limit the price premiums sustainable by existing market participants, by increasing actual or potential competition
- by shortening the supply chain, distribution and inventory cost will be lower.

Estimates of NAIRU suggest that the low level of unemployment over the past few years should have produced a fairly significant acceleration in prices, yet inflation has continued to decline. Some, like Robert Gordon (1997) and Staiger, Stock, and Watson (1997), take this occurrence as evidence that the NAIRU has declined. Others argue that special factors, such as movements of employee health coverage to health maintenance organizations, have temporarily masked the increase in inflation. Another often cited explanation for the surprisingly good inflation performance of the late 1990s concerns the increasing sensitivity of the US economy to foreign economic conditions. Since capacity utilization abroad has been slack in recent years, it is argued, US inflation has remained mild. This is due, it is further argued, to the globalization of the world economy, to which we now turn.

Globalization of the world economy

'Globalization' in its economic aspect refers to the increasing integration of economies around the world, particularly through trade and financial flows. The term sometimes also refers to the movement of people (labour) and knowledge (technology) across international borders.

Increasing and unprecedented globalization, driven partly by technological change and partly by the deliberate removal of government-created barriers to the international movement of goods, services, people, financial capital, enterprises and ideas, has transformed the international and domestic competitive environments.

At its most basic, there is nothing mysterious about globalization. The term has come into common usage since the 1980s, reflecting technological advances that have made it easier and quicker to complete international transactions – both trade and financial flows. It refers to an extension beyond national borders of the same market forces that have operated for centuries at all levels of human economic activity – village markets, urban industries or financial centres.

Markets promote efficiency through competition and the division of labour – the specialization that allows people and economies to focus on what they do best. Global markets offer greater opportunity for people to tap into more and larger markets around the world. It means that they can have access to more capital flows, technology, cheaper imports and larger export markets.

Critics of the view that globalization can raise productivity and contain inflation, particularly Krugman (1997), point out that about 85% of the US economy, primarily services, is not subject to the intense pressure of the market place.

The New Economy and technology stocks

The ten essential principles for doing business in the New Economy

In an oft-cited article in *Business 2.0*, a New Economy magazine that has now ceased publication, the principles of what were perceived to be the ten driving forces behind the New Economy were set out. As these principles had a significant impact on the valuation of technology/Internet stocks, and are discussed further throughout the text, they are set out below.

1. MATTER: It matters less

It's a cliché, but it's the key to the New Economy: processing information is dramatically more powerful and cost-effective than moving physical products. Increasingly, the value of a company is to be found not in its tangible assets but in intangibles: people, ideas and the strategic aggregation of key information-driven assets. (See Chapter 5, where we discuss these ideas in more detail.)

2. SPACE: Distance has vanished. The world is your customer and your competitor

Geography has always played a key role in determining who competed with whom. Now your business can connect instantly with customers all over the globe. Flipside: you're exposed to world-wide competitors as well. The opportunity – and the threat – has never been greater. (See Chapter 5 again for more discussion of this idea.)

3. TIME: It's collapsing. Instant interactivity is critical, and is breeding accelerated change

In a world of instantaneous connection, there is a huge premium on instant response and the ability to learn from and adapt to the market place in real time. Winning companies accept a culture of constant change, and are willing to constantly break down and reconstruct their products and processes – even the most successful ones.

4. PEOPLE: They're the crown jewels . . . and they know it

Brainpower can't be tallied on a ledger sheet, but it is the prime factor driving the New Economy. More than ever before, huge value is being leveraged from smart ideas – and the winning technology and business models they create. So the people who can deliver them are becoming invaluable, and methods of employing and managing them are being transformed.

5. GROWTH: It's accelerated by the network

The Internet can dramatically boost the adoption of a product or service by 'viral marketing', network-enhanced word of mouth. Communication is so easy on the Web; product awareness spreads like wildfire. So once a company reaches critical mass, it can experience increasing returns leading to explosive growth. This principle means that in the New Economy, first-mover advantages are greater than ever.

6. VALUE: It rises exponentially with market share

For products that help to establish a platform or a standard, the network effect is even more pronounced. The more plentiful they become the more essential each individual unit is, a striking exception to the economic rule that value comes from scarcity. In addition, some companies give away their products to establish market share, then sell linked services later on: network effects were experienced historically in the adoption of telephones and fax machines. The difference today is that everyone is linked, so far more products and services gain their value from widespread network acceptance, as we discussed earlier in this chapter.

7. EFFICIENCY: The middleman lives. 'Infomediaries' replace intermediaries

Traditional distributors and agents are seriously threatened by a networked economy in which buyers can deal directly with sellers. But a new brand of middleman is being created. As the amount of info-clutter grows, these infomediaries are needed to turn dumb data into usable information. They offer aggregated service, or intelligent customer assistance, or powerful technology-based buying aids, or an attractive, community-based buying environment.

8. MARKETS: Buyers are gaining new power – and sellers new opportunity

It's no longer necessary for your customer to walk down the street to compare prices and services. Your competitor may just be a mouse-click away. And intelligent software helps buyers find the best deal. So businesses that genuinely offer unique services or lower costs will flourish, benefiting from a flood of new buyers. Those that have relied on physical barriers to competition will fail.

9. TRANSACTIONS: It's a one-on-one game

Information is easier to customize than hard goods. The information portion of any good or service is becoming a larger part of it.

10. IMPULSE: Every product is available everywhere. The gap between desire and purchase has closed

The shelf space of the World Wide Web is unlike any other, in that it has no bounds. Artificial constraints on choice are replaced by the ability to purchase the precise product you desire. The impulse to buy and the purchase itself used to be separated by a combination of physical and mental barriers. When you heard a song on the radio, you had both to remember the song or the artist and actually go to a store to purchase. Online it's different. Discover a product you desire, and just hit the 'buy' button. Consequence: The processes for marketing, sales and fulfilment are merging.

Postscript

Although the magazine promoting these ideas has since ceased publication and we have had, of course, a Nasdaq stock price collapse, there is little doubt that these ideas have not gone away. They may have been temporarily set back but they will return. They play a key role in understanding the issues associated with valuing technology/Internet stocks.

But let us now turn to the micro-economic picture. As discussed earlier in this chapter, the New Economy ideas are indeed still powerful. But to what extent are they reflected in how we measure technological change?

Technological change and US economic statistics

The US economy is constantly evolving. New technologies are continuously transforming the production and delivery of new and existing goods and services. But business practices adapt just as readily to changes in technology. The scramble for market share and profitability forces firms to find innovative ways to add value to their firm's products and services.

These innovations often require computer hardware and software. In the past, when businesses generally used mainframe computers, they acquired the hardware and software as a bundle. With the advent of personal computers and minicomputers, a greater percentage of business software was purchased separately from hardware. Yet business software still continued to count as an investment in the US National Income and Product Accounts (NIPA) only when purchased as installed software on a new computer. Software purchased separately was considered an intermediate input and did not count as capital investment. Hence, the unbundling of computer hardware and software purchases led to the unintended consequence of US statisticians classifying a considerable portion of business software expenditures as intermediate products, and hence not directly raising the standard measurement of economic activity, GDP.

Given the mid to late 1990s' boom in computer purchases, and the fact that computer software, like other capital expenditure, provides a flow of services that last more than a year, the US Bureau of Economic Analysis (BEA) recognized the need to address this issue, and did so in the 28 October 1999 release of the advance third quarter GDP report. Accompanying the advance report was the 11th comprehensive revision of the NIPA. The BEA now treats business purchases of computer software and 'in-house' software production as fixed investments in order to remedy the classification problem discussed above.

Adding business software purchases to the new equipment and software (E&S) component of non-residential fixed investment, formally known as producers' durable equipment (PDE, raises the level of GDP. In 1998, for example, the nominal value of US software investment totalled $123.4 billion, which was 15.1% of E&S investment and 1.4% of GDP. A back-of-the-envelope calculation shows that reclassifying software purchases as fixed investment boosted the growth of real GDP by about 0.1 percentage point a year during the recent US business expansion (first quarter of 1991 to first quarter 2001). Moreover, with software prices falling much less than prices of computers and peripheral equipment during this period (1.6% versus 19.3%), the inclusion of software prices has had the effect of slowing the rate of decline of prices of information processing equipment (computers, software and other equipment).

The effect of these statistical revisions is that software purchases now boost the GDP growth figures. Previously when corporations and governments bought software the Bureau of Economic Analysis (BEA) rated it as 'purchased input', much as when an auto company bought steel or a newspaper bought newsprint. So now software has been added to the 'business investment' column meaning that the BEA recognizes that software lasts more than a year, adding directly to the computed value of GDP.

So the New Economy driving principles have now been finally incorporated into the official statistics. Given this fact, are there any lessons we can learn from technical progress from a historical perspective? In particular, what can we learn from the works of Joseph Alois Schumpeter and a more contemporary follower, Alvin Toffler?

Schumpeter: the pioneer of the analysis of technological change

Austrian born economist Joseph Alois Schumpeter (1883–1950) viewed the business cycle as a continuing process moving from phase to phase as economic development occurs. Schumpeter located the cause of long run economic expansions in the investment opportunities provided by a major technological breakthrough. More particularly, Schumpeter distinguished between 'inventions' and 'innovations'. He defined inventions as the discovery of new ways of producing. These occur more or less continuously in modern market economies as people are always discovering new ways of producing or better ways of producing. However, inventions become economically significant only when they are actually introduced into economic activity. This he called 'innovation' and Schumpeter argued that such is the nature of modern economies that innovations occur only discontinuously. They tend to be bunched or clustered and this cluster of investment opportunities, exploited more or less at the same time, produces an expansion.

In general, innovation for Schumpeter meant a change in production severe enough to cause a discontinuity in production. The most basic kind would be the introduction of a new product with a new production process, particularly one requiring the construction of a new plant or equipment to carry it out. The introduction of the production line to produce Henry Ford's new automobiles is the clearest historical example of an innovating entrepreneur. There have been other kinds of innovation involving new sources of raw materials, new markets for products, or new methods of organization in an industry. But whatever the type, the distinctive character of the innovation for Schumpeter was that it required introducing fresh ideas into the production process and these involved investment outlays to bring them into being. Schumpeter argued that to innovate successfully an entrepreneur must have 'broader horizons', his term for the range of new possibilities that any entrepreneur felt comfortable contemplating.

Significant innovations were therefore associated with the rise to leadership of new entrepreneurs. These infrequent leadership changes occurred only when dissatisfaction with old ways reached a critical level. This level was associated with the building up of tension in the firm caused by slow or non-existent progress under old leadership, and the consequent willingness of those who decide on entrepreneurial fates to take a chance on new leaders.

Innovation in one industry or firm tended to result in other firms taking a chance in what Schumpeter called 'imitation waves'. The introduction of new techniques and inventions was thus bunched – the cluster of innovations about which Schumpeter wrote: 'Progress – in the industrial as well as in any other sector of cultural or social life – not only proceeds by jerks and rushes but also by one-sided rushes productive of consequences other than those which would ensue in the case of co-ordinated rushes. In every span of historic time it is easy to locate the ignition of the process and to associate it with certain industries, with certain firms, from which the disturbances spread over the entire system' (Schumpeter, 1939).

Schumpeter, therefore, developed a broad theory to account for instability in modern industrialized market-oriented economics. This theory focuses on the profit-motivated activity of entrepreneurs who operate within a given culture and are heavily influenced by all the experiences that shape as well as limit their perspective and vision.

The 2000 Nasdaq technology stock price collapse is symptomatic of the fact that there has no doubt been massive misallocation of resources in the US economy, particularly over-investment in technology. This creative destruction will be painful. Schumpeter would have surely welcomed the sight of the turbulence and creative destruction associated with the volatility created by the New Economy.

Tofler's 'third wave'

In Alvin Toffler's 1980 book *The Third Wave*, based partially it must be said on earlier work by Schumpeter, the wave metaphor is applied to symbolize the fact that beneath the raging surface of change it is essential to distinguish those changes that are merely transitional and cosmetic from those that are truly revolutionary. Despite the fact that Toffler's work was published over 20 year ago, its impact is still being felt and is worth highlighting here.

The first wave was the change from a hunter–gatherer existence to a more settled agricultural way of life. The second wave was the industrialization that occurred in (mainly) Europe and North America between the eighteenth and mid-twentieth centuries.

Second wave society was heavily dependent on concentrated fossil fuels for its energy sources, factory production for its goods, the nuclear family for social stability, the corporation as the main means of wealth creation and mass media to supply the labour for the factories. Mass education and the mass media contributed to social cohesion (and to social engineering). There was a wide gulf between production and consumption: almost no one produced the items they consumed. The whole organization was managed by a set of elites – governments, civil servant and corporations – whose task was to integrate the elements into a whole. In particular, the nation-state was a second wave creation, particularly matched to the size and scope of second wave economies.

The forces of the second wave are illustrated in Table 2.1.

According to Toffler, the characteristics of the third wave are almost diametrically opposed to those of his second wave shown

Table 2.1 The forces of the second wave

Force	*Consequences*
Standardization	Standardization of measurements; millions of identical products
Specialization	Specialists within manufacturing and in the professions
Synchronization	Groups of workers working together at the same time
Concentration	Concentrated energy sources, cities, factories, schools, hospitals, mental asylums and of capital
Maximization	Quest for efficiency and bigness
Centralization	New management methods in business and politics

Source: Toffler, 1980

in Table 2.1. Although standardization of measurement in production has been retained, there has been a move away from the production of millions of identical products to providing much greater choice. For example, UK supermarkets stock around half a dozen types of table salt; there are around 250 models of car on sale in most European countries. There has been a demassification of the media, as shown by the upsurge in the number of specialist periodicals and the number of TV channels now available. And whereas in the second wave the recipients of goods and services were passive, having a negligible individual impact on the offers they bought, the move now has been towards incorporating individual tastes into their production.

Specialization was a very significant feature of the second wave. Instead of an individual or their family taking responsibility for their own food, clothing, education, health, etc., large numbers of specialisms became established by people who would do this for them: specialist farmers, teachers, doctors, lawyers, etc.

In the third wave, in contrast, individuals are taking back some of these responsibilities. In the third wave information technology and knowledge plays a key role. This latest wave of human endeavour, updating Toffler, can be referred to as the 'information age or knowledge age', a theme attention is drawn to throughout this text, given its importance for valuing technology stocks.

The macro-economic consequences of the Internet economy

Apart from the individual winners and losers, the Internet may create real efficiencies for the economy as a whole. For some guidance as to where these efficiencies might come, consider the driving forces behind the past 20 years of financial market innovation. Nobel Prize-winning economist Robert Merton identifies the source of the efficiencies in three places.

The first is in the creation of new and bigger markets, which can lead to lower prices and higher quality because firms can now complete transactions with firms that were either too far away or in markets that were too small to be profitable. The second is reducing the costs of making those transactions. The third is a bit subtler, but arises because each party to any business deal usually has inside information about his or her own company. This asymmetrical knowledge tends to reduce trust between the parties and increase the cost of monitoring subsequent performance. If the Web really were to promote the sharing of information and to make the boundaries between firms less distinct, it would shrink the costs that are driven by distrust. As stressed throughout this text, we do indeed live in an information age.

3

Technology and the Internet economy: what made it all happen?

The story of the Internet and the World Wide Web revolves around one man, Tim Berners Lee. There are three defining features of the World Wide Web and we come back to these later in this chapter. These are:

- the Uniform Resource Locator (URL), which forms its addresses;
- the Hypertext Transfer Protocol (HTTP), which allows hyper-text to be transported over networks;
- the Hypertext Markup Language (HTML), which is the language of the Web documents. It is HTML that allows for the easy linking of files and pages by embedding an inter-active reference to a document directly into the page.

All three defining features of the Web were in Tim Berners Lee's prototype browser/editor designed at CERN, the European Particle Physics Laboratory in Switzerland, in the late 1980s. Many of these ideas are, of course, simply milestones in the development of information technology. This chapter focuses mainly on the evolution of the Internet economy. The Appendix at the end of the chapter offers a broader historical perspective on the milestones in information technology.

What is the Internet?

The term Internet is a shortened version of 'international net-work of computers'. The Internet is the infrastructure linking the computers of the world together in a network of networks consisting of links and nodes. A national road system is an analogous network of networks. In any city there are roads (the links) and roundabouts and crossroads (the nodes) forming a network. Each city network is linked to other city networks via highways. The networks of the Internet operate under the Transmission Control Protocol/International Protocol (TCP/IP), a standard protocol that is recognized by virtually all computers.

It is often asked whether the Internet is the same as the World Wide Web? The answer is quite emphatically No! Often the term World Wide Web (WWW), or simply the Web, is used synonymously with the Internet, but Web applications are only one category of application where data are transmitted across the Internet; others are e-mail, voicemail and electronic data interchange (EDI). Although e-mail, voicemail and EDI can and are run on the Internet *outside* of the Web – and indeed outside of the Internet – rather confusingly perhaps, they are also available *within* the Web.

What is so unusual about the Internet is that since it has become available to the general public, its penetration has been exceedingly fast. As the US Department of Commerce (1995) commented: 'It took 38 years for the radio to reach 50 million US listeners, and 13 years for the television to reach 50 million viewers. It has taken only 5 years since the Internet got going for real in the US in 1993 for the Internet to acquire 50 million users.' Thinking about this quote, it's worth noting the increase in activity possible to the recipients: first they were listeners, then listeners and viewers and now *users*.

What is the Internet economy?

The Internet economy is made up of companies that directly generate all or part of their revenues from Internet-related products and services. This is a two-tier economy consisting of infrastructure and economic activity (see Figure 3.1).

The infrastructure can be broken down further into two components:

(1) The physical Internet infrastructure of e-commerce (Layer 1)
- Fibre-optics makers: Corning, Pirelli
- Networking hardware/software: Cisco, Lucent, 3Com

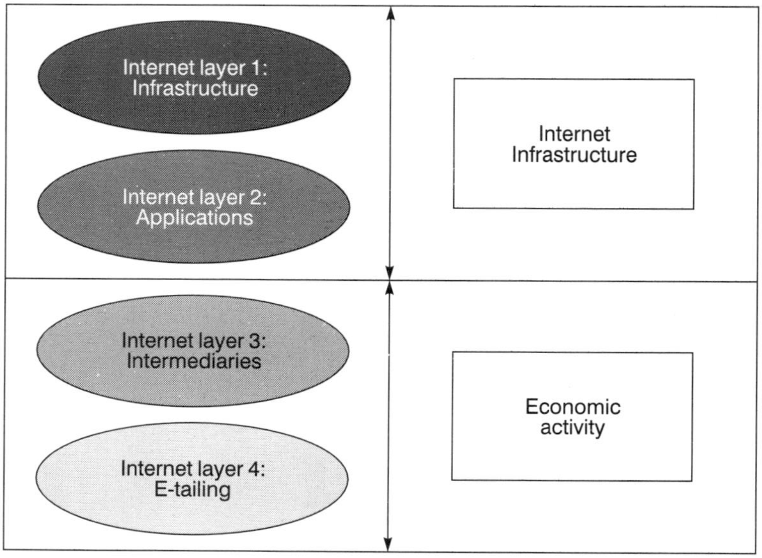

Figure 3.1 The Internet Economy

- PC and server manufacturers: Dell, Compaq, Intel
- Internet service providers (ISP): AOL, Earthlink
- Security vendors: Verisign, Entrust
- Internet backbone providers: MCI Worldcom, Qwest

(2) Internet applications (Layer 2)

- Web development software: Adobe, Vignette
- Web-enabled databases: Oracle, IBM, MS SQL Server
- Search engine software: Inktomi, Verity
- Internet commerce applications: Netscape, Microsoft, Sun, IBM
- Transaction processing companies: First Data, ADP
- Web hosting and support services: Exodus, Globix, Verio

Economic activity, as shown in Figure 3.1, is also broken into two components:

(3) Internet intermediaries (Layer 3)
Internet intermediaries increase the efficiency of electronic markets by bringing together buyers and sellers over the Net and facilitating interaction between them.
   (a) On-line search, evaluation, communication, co-ordination and assurance of vendor and product/service quality are the most important aspects of the Internet economy.

(b) Intermediaries fill a crucial gap in information and knowledge, which would otherwise impair the functioning of the Internet as a business channel. Examples here would be:
- Internet portals: Yahoo, Excite
- Internet AD brokers: Doubleclick, 24/7 media
- Vertical market makers: Verticalnet, PCOrder
- Content aggregators: Cnet, Zdnet, Broadcast.com
- On-line auction: QXL, Ebay
- On-line brokerage: E*Trade, Schwab.com

(4) On-line transactions (Layer 4)
- E-tailers: Amazon,
- Manufacturers selling on-line: Cisco, Dell, IBM
- On-line travel: Travelweb, 1Travel
- Subscription-based companies: TheStreet.com, Wired
- On-line entertainment: Disney.com, ElectronicArt.com

The above Internet economy structure is very simplistic in its form but it is important to remember that, like most other industries, there are players who span various layers. The layer descriptions show Cisco and Dell in Layer 1 as well as Layer 4. AOL are also a key player in Layers 2, 3 and 4, and with the Internet culture the ability to diversify into new products and services will mean more revenue from a bigger brand presence and also the potential target markets from their present services. Other players will no doubt span several layers.

What is the World Wide Web?

The World Wide Web is like an encyclopaedia, a telephone directory, a record collection, a video shop and Speaker's corner all rolled into one and accessible through any computer. It has become so successful that to many it is synonymous with the Internet; but really the two are quite different. The Internet is like a network of electronic roads criss-crossing the planet – the much-hyped 'information superhighway'. The Web is only one of many services using that network, just as many different kinds of vehicle use the roads. The Web just happens to be by far the most popular. The arrival of the Web in 1990 was to the Internet like the arrival of the internal combustion engine to the country lane. Internet transport would never be the same again. But how was the Internet able to evolve into the World Wide Web?

The telephone network

Until the arrival of computer networking, the telephone system was the only medium for electronic mass communication. Although the two are intimately linked, there are important differences between the way people talk to each other and the way computers converse. In a telephone conversation, people tend to talk continuously, or even both at once. For that reason, the electronic link between them is reserved exclusively for them and remains open for as long as they wish to speak. Computer conversation, on the other hand, is a rather more staccato affair. It comes in fits and starts, so keeping a link open continuously would be extremely wasteful.

Communicating by fax is a bit closer to computer communication. Information is bundled up into discrete packages and a line is held open just long enough for the package to be sent. Once the message has been received, the connection is closed and the message can be digested 'off-line'.

When you pick up the telephone and dial a number, the telephone system finds you a line to the person you want to talk to. The problem with such a system is scalability: it is hard to make the system grow larger. Each new subscriber would have to be linked to each existing subscriber. Such a system would not scale up easily to more than a handful of subscribers.

The solution is circuit switching. Instead of every subscriber having a line to every other, each one has a single line to a central switch. It works because most people are off the phone most of the time and only a fraction of all possible pairs of phones are connected to each other at any given moment. In the old days there would be an army of switchboard operators connecting calls. Now it is all done automatically; but the basic principle is the same.

A circuit-switched system makes the scalability problem more tractable. Each community has its own local switch and different communities are linked by long-distance connections between switches. As cables gave way to optical fibres, which can carry much more traffic, the efficiency of communication improved.

Circuit switching links you to the person you are calling by establishing a circuit between the two of you. That circuit may be made from several pieces of wire and lengths of optical fibre, but as long as you want it, it is exclusively yours. You are also paying for it even when you are not saying a word, and no one else can use it as long as you are on the line.

The capacity of a circuit is called its bandwidth: it is the amount of information the circuit can carry per second. Information, like a human voice, doesn't take up much bandwidth. High-fidelity stereo takes up nearly thirteen times as much bandwidth, and television-quality pictures need nearly 30 000 times the bandwidth of speech.

The quality and length of a cable determines the bandwidth it can carry, but the bandwidth of a circuit is no better than that of its weakest link. Optical fibres can, at the latest count, carry a massive four million conversations at once.

The Internet and protocols

Every time we make a phone call or meet someone in the street, we use a set of protocols to conduct our communication. When we bump into someone we know, we say, 'Hello'. We might shake hands, enquire about our acquaintance's state of health. Computer communication also relies on protocols.

The Internet is a collection of computer networks talking to each other using what is called packet switching. All communication between computers on the Internet happens by cutting things up into small packets and sending them through a system of electronic routing stations to their destinations. Imagine a computer network made up of computers A to F connected though routers 1 to 5, then a message from A to E might pass through routers 1,4 and 3, but if 4 broke down the message could still pass through 2 and 5. Even if the first packet of a message went via 4 before it broke down, the remaining packets could take different routes and E would still be able to reassemble them into a coherent message. Each router would contain a routing table telling it which way to pass on the message, with a back-up pass-on address in case the first didn't work.

If all computer networks had been developed to a single standard, there would be no Internet, just one big, ever-expanding network. But that is not what happened. Many different packet-switching networks using many different protocols have been developed, and the Internet is the result of connecting them all together.

The basic set of protocols at work on the Internet is called TCP/IP. IP, the Internet Protocol, is the lingua franca of computer communication: it is the protocol that routers use to pass packets on. Individual networks may use protocols other than TCP/IP, but the router connecting them to the Internet, known as a gateway, must hand packets over using IP. TCP,

Transfer Control Protocol, is the protocol at work in the sending and receiving computers. Its job is to break information up into packets, each one properly labelled with the sender's and receiver's address, to ensure that they all arrive, and to reassemble them when they do. If the IP loses a packet, it is TCP's job to ask that it be resent.

The defining features of the World Wide Web

The three defining features of the World Wide Web are the Uniform Resource Locators (URLs) that form its addresses, the Hypertext Transfer Protocol (HTTP) that allows hypertext to be transported over networks, and the Hypertext Markup Language (HTML) that is the language of Web documents. These were all already there in Tim Berners Lee's prototype browser/editor, although as a user you didn't have to know about any of them. Creating a Web document was as simple as typing. The browser/editor took care of arranging what you had written into HTML for other browsers to interpret. URLs weren't displayed: if you really wanted to see one, or type one in explicitly, you had to open a special window. The idea was that using the Web would be as close as possible to people's everyday experience of computers, and they wouldn't have to learn anything new. But inside the URL was hidden all the power of the Web.

The URL built on the Domain Name System (DNS) invented nearly two decades before by Paul Mockapetris, and because of that, it was scalable. It was Paul Mockapetris who, when he invented the DNS, created the format for e-mail addresses so that they look like bkettell@hotmail.com (addressing by organization/domain) rather than bkettell@f.isi.edu (addressing by machine name) No matter how big the Web became, there would always be enough URLs to go round.

What is an Internet Stock?

Before one can appreciate the valuation issues involved in the Internet economy, a brief description of an Internet stock is essential. The term Internet stocks refers to stocks in a company classifiable under the following three categories:

- Pure Internet company: The high profile companies such as Amazon.com, eBay, Etrade and Yahoo are referred to as

'pure' Internet companies. These companies did not exist before the Internet, their whole business model is built around the Internet, and all their revenues derive from transactions on the Internet.
- Bricks-and-mortar companies: These are traditional companies that are successfully changing their business model to seize the opportunities offered by the Internet. Examples include companies such as Federal Express and Disney. Federal Express has totally redesigned its business processes so that 60% of its shipments are fully ordered, tracked and managed via the Internet.
- Internet technology companies: These companies make the switches, routers, modems, software and other technology critical to the Internet's operation. A good example of such a company is Cisco Systems.

Overall, Internet stocks can be stocks in a company that falls within any of the above three categories. However, the main bulk of the recent Internet stock craze took place with stocks that fell under the 'pure' category, and throughout this book reference to an Internet stock will normally be referring to a stock in a company that can be described as a 'pure' Internet company. There are of course many exceptions, with companies that occur in the other two categories.

Developments in information technology

The transistor, the basis for modern computers, was discovered in 1947. Computing and communications have developed enormously over the subsequent 50-plus years. The increase in computing power – measured over three decades – followed Moore's law, named after Gordon Moore, co-founder of Intel, the world's leading maker of computer chips. In 1965 Moore forecast that computing power would double every 18 months to 2 years. This meant, for example, that between 1960 and 1990 the cost of a unit of computing fell by 99%. Computer power is now 8000 times less expensive than it was 30 years ago. As Bill Gates, the founder of Microsoft famously commented, 'If we had similar progress in automotive technology, today you could buy a Lexus for about $2, it would travel at the speed of sound and travel for about 600 miles on a thimble-full of gas.'

Computing technology has been one of the main drivers of the convergence; the other is the rapid development of fibre-optic and wireless technologies, both satellite and land-based. While computer technology provides processing power and the means of routing signals along the appropriate communication 'roads', these roads are increasingly fibre-optic and wireless. Fibre-optic cables are made from filaments of glass fibres, each only the thickness of a human hair and from glass that is so pure that a sheet 70 miles thick would be as clear as a windowpane. The filaments carry the bits as very small pulses of light. Fibre-optic and satellite channels support the revolution in communications because they provide what is termed broad bandwidth. Bandwidth refers to the number of bits per second (bps) that can pass through a channel.

There is a comparable law to Moore's for fibre-optics – currently, bandwidth is tripling every year. This is known as Gilder's law and is discussed further below. In addition, there is Metcalfe's law, promulgated by Robert Metcalfe, who founded 3Com, the US networking equipment manufacturer: 'The value of a network is proportional to the square of the number of nodes in the network', again discussed below.

Moore's law: the processing power of microchips

Data transmission costs and computing costs are plummeting. Why is this important?

One of the chief enablers of the New Economy is instantaneous global communications: the ability to easily send and receive data – everything from documents to video and multimedia – inexpensively. One measure of progress in that direction is the cost of data transmission. The cost to transmit one bit of data over a kilometre of fibre-optic cable declined by three orders of magnitude between the mid-1970s and the beginning of the 1990s, allowing more data to be transmitted over longer distances at lower prices.

Technologies for transmitting data are also getting more and more powerful. For example, technology recently developed by Lucent transmits 3.2 terabits – which is approximately equal to 90 000 volumes of an encyclopaedia – per second.

Computing costs continue to fall. In a 1965 speech, Intel chairman (now emeritus) Gordon Moore made a famous observation: With price kept constant, the processing power of

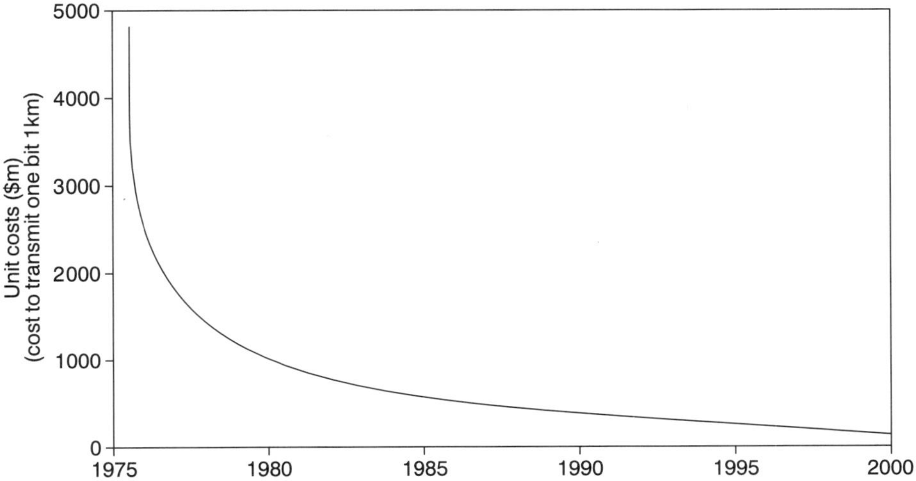

Figure 3.2 Moore's law and data transmission costs

microchips doubles every 18 months. Moore's law, as it came to be known, and is illustrated in Figure 3.2, governs Silicon Valley's most important product cycles, for hardware and software alike. And its relentless drive has spawned whole new industries: digital watches, calculators, videogames, the Internet – with smart phones, digital TV, etc.

In fact Moore's law, which says that the processing power of microchips doubles every 18 months, has a corollary: the cost of computing is dropping by nearly 25% per year. In 1978, Intel Corp introduced its 8086 chip, which defined the base architecture for the later x86 series (including the 386, 486 and Pentium chips). It contained 29 000 transistors. Four years later came the 286, with 134 000 transistors. Three years after that, the 386 had 275 000 transistors. And on the trend goes: the Pentium Pro, introduced in 1995, had 5.5 million transistors in its core central processing unit. Meanwhile, the cost of all that computing power has been dropping precipitously. In 1978, the price of Intel's 8086 was 1.2 cents per transistor, and $480 per million instructions per second (mips). By 1985, the 386 cost 0.11 cents per transistor and $50 per mips. Ten years later, the Pentium Pro's introductory price amounted to 0.02 cents per transistor, and $4 per mips.

In September 2001 Intel announced that it was selling a Pentium 4 processor that runs at two gigahertz, or two billion cycles per second, again marking the doubling of computer chip speed in only one year and a half (see Figure 3.3). The prices will no doubt continue to fall in the future.

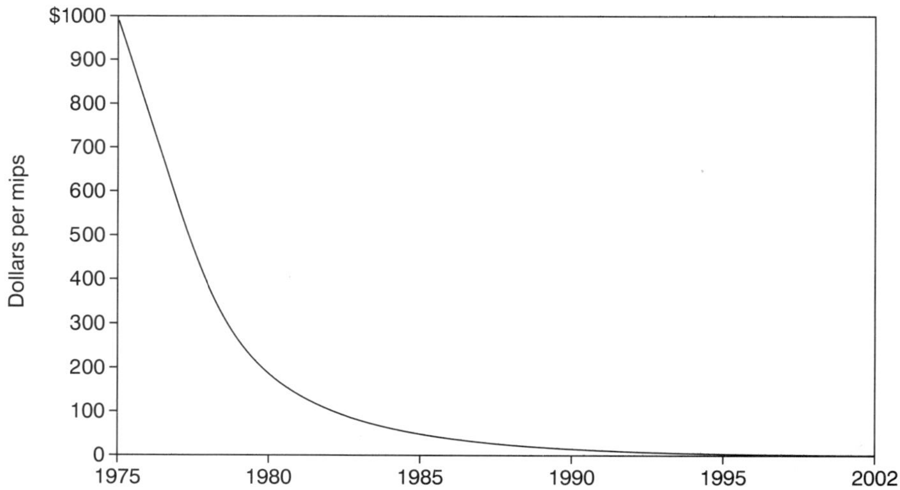

Figure 3.3 Microprocessor price trends

Gilder's law: developments in bandwidth

The term bandwidth is used to mean the slice of the radio spectrum available for transmission. Today it is mostly used to describe the rate at which information – measured in bits of data per second – can move between computers.

The ability to transfer large amounts of data is largely determined by bandwidth, the carrying capacity of the connections, or the 'size of the pipes', between the sender and the receiver of the data. Greater bandwidth allows faster transmission of larger amounts of data, which in turn facilitates not only the development of vastly more valuable and compelling on-line services, but also the convergence of all forms of electronic data transmission, from e-mail and basic text documents, which require relatively little bandwidth, to full-motion, real-time video applications, which require a great deal of carrying capacity.

Fibre-optic cable, currently being laid as fast as the companies supplying it can dig trenches, are increasing this carrying capacity. Pundit George Gilder has proposed a bandwidth corollary of Moore's law: Backbone capacity will triple annually for the next quarter century. It could happen. Already, corporate Internet users are measuring their access in gigabits per second – sufficient to start realizing the trillion-dollar pipe dream of TV and Internet convergence.

Metcalfe's law: the impact of network externalities

In June 1973 Bob Metcalfe, working with Vincent Cerf and Robert Kahn, who in 1974, as mentioned earlier, published the blueprint for the Internet, set out what became known as Metcalfe's law. This states that 'the value of a network is proportional to the square of the number of nodes in the network'. The basic idea is that after a certain point the value you get from connecting computers and devices outstrips the cost.

These benefits generate what economists call network externalities. Fax machines and board games share an economic quirk, each new one sold – or, in the case of games, each new enthusiast who learns the rules – adds to the value of the rest. A fax machine isn't worth much if there are no others to communicate with; board games aren't much fun if no one else knows how to play. Thus the whole adds to the value of each of its parts.

Network externalities – a term for the effect one person's decision to buy into a network has on others who are still thinking of buying in – have been the Net's rocket fuel: the more people who connect, the more valuable a connection becomes. But the Internet in turn is bringing network externalities to the economy as a whole, a concept we discussed earlier in Chapter 2 when examining some of the aspects of the New Economy. Knowledge is affected by the same sorts of network externalities as Internet connections themselves; having the equipment to receive messages is no more important than having the knowledge to understand them. This explains why the future seems to happen so fast on the Internet. Change accelerates itself. And yesterday's arcane knowledge becomes today's essential information.

So Metcalfe's law, that the value of a network increases as the square of the number of linked devices, creates an environment in which network externalities guarantee the benefits of increased Internet usage.

Appendix 3.1

Technology milestones in the evolution of the Internet

1447	German printer Johann Gutenberg invents movable type.
1452–1519	Life of Florentine artist and scientist Leonardo da Vinci, who makes discoveries in medicine, hydraulics, meteorology and aerodynamics.
1530	In a treatise, Polish astronomer Nicolaus Copernicus postulates that the Earth revolves around the sun. Despite church protest, the work lays the foundation for scientific thought governed by logic and deductive reasoning.
1600	English physician William Gilbert discovers electrical and magnetic properties.
1608	Jans Lippershey, a Dutch spectacles maker, invents the telescope.
1564–1642	Life of Italian physicist Galileo Galilei, who discovers the laws of falling bodies, reportedly by throwing heavy objects from the Leaning Tower of Pisa. In 1609, drawing on Lippershey's work, he invents a telescope of 20× magnification. He embraces Copernicus's theory of the universe, as this is the only one that will explain his observations of tide movement. For this, he is condemned to life imprisonment for heresy, commuted to house arrest by the Roman Catholic Church. In 1992 the Vatican acknowledges its error and reverses the decision.
1827	German physicist George Simon Ohm develops the law of electric resistance.
1825–1840	Charles Babbage, a British mathematician, creates plans for calculating devices that are capable of using stored programs as well as memory stored on punched cards. Unable to find funding, he never built the machines. The first one, the Difference Engine, was finally built in 1991 and performed flawlessly.
1833–1842	Working with Babbage was Ada, Countess of Lovelace and daughter of Lord Byron. Working with his designs, she developed the concepts of

	modern programming. In 1979, the US Department of Defense named a programming language 'Ada' in her honour.
1837	American inventor and artist Samuel Morse invents the magnetic telegraph and Morse code.
1847–1931	Life of Thomas Alva Edison, who came up with more than 1000 inventions, including the electric light bulb, the electric generator, the stock ticker and the phonograph.
1854	English mathematician George Boule develops Boolean, or binary, algebra, which is instrumental in the development of computer systems.
1876	Alexander Graham Bell, who began his career as a teacher for the deaf, invented the first workable telephone and microphone.
1888	George Eastman patents the box camera, the first to use roll film. It sells for $25.
1895	Guglielmo Marconi created wireless telegraphy when he sent a signal a few kilometres. In 1901, he sent a radio signal from Cornwall, England to St John's, Newfoundland; he won the Nobel Prize for physics in 1909.
1898	Danish Engineer Valdeman Poulsen invents the magnetic tape recorder.
1905	$e = mc^2$, concludes physicist Albert Einstein. The theoretician of relativity wins the Nobel Prize for physics in 1921.
1923	First transatlantic telephone call connects New York and Britain.
1925	Phototelegraphy is invented at Bell Laboratories.
1936	British mathematician Alan Mathison Turing wrote a paper 'On the Theory of Computable Numbers', devising theoretical plans for a machine capable of performing any mathematical calculation. His work was important in later development of the digital computer. He laid the foundation for artificial intelligence, developing the Turing Test to determine whether a computer could 'learn' or 'think'.
1938	Chester Carlson invents xerography. He tried unsuccessfully to sell the process to more than 20 companies. The Haloid Co. purchases the rights in 1947 and changes its name to Xerox.

1939	Stanford engineers William Hewlett and David Packard start a company on $500, and set up shop in a garage in Palo Alto, California. A coin flip determines whose name goes first.
1940	Radio Detection and Ranging, or Radar, is developed by Scottish physicist Robert Watson-Watt.
	Konrad Zuse of Berlin invents the first programmable calculator, called the V1.
1941	Iowa State College professor John Vincent Atanasoff and graduate student Clifford Berry develop the first digital computer, which they called the ABC. John W. Mauchly and colleague J. Prepser Eckert later patent a digital-computer design, forming a company that later becomes Sperry-Rand Corp.
1943	Harvard scientist Howard Aiken develops the first program-controlled computer, the Mark 1, independently of Zuse. It is 15.3 metres long and weighs 4.5 metric tons. The US navy uses it to calculate missile trajectories.
	Eckert and Mauchly develop the Eniac, or electronic numerical integrator and computer. It weighs 27 tons and has 18 000 vacuum tubes. The US army buys it to calculate trajectories and to develop the hydrogen bomb.
1947	Bell Labs makes the first cellular telephone.
1948	The transistor is invented at Bell Labs by William Bradford Shockley, Walter Houser Brattain and John Bardeen, for which they win the 1956 Nobel Prize for physics.
1950	Eckert and Mauchly develop the first commercially available computer, the UNIVAC-1. This is the first general-purpose computer, capable of using both numbers and text.
1952	Grace Murray Hopper, a navy programmer working in Howard Aiken's Harvard lab, devises the first compiler, which translates English-like programming code into the ones and zeros of machine language. In 1957 she develops the Common Business-Oriented Language, or Cobol. She also coined computer terms: Investigating a problem with a Mark 1 vacuum tube, she pulled out a moth and, legend has it, exclaimed, 'It's a bug!'.

1954–58	Formula Translation, or Fortran, the first high-level programming language, is developed by Jim Backus and others at IBM.
1956	The first transistorized computer is developed at MIT.
1957	Light amplification by stimulated emission of radiation, known as laser, springs from the research of American Gordon Gould. Sputnik, the first satellite, was launched by the Soviet Union on 4 October. It spurred a revolution of science education in the US.
1959	Jack Kilby and Robert Noyce of Texas Instruments develop the integrated circuit. Noyce goes on to found Intel Corp.
1963	The first minicomputer, the PDP-8, is developed at Digital Equipment Corp. Douglas Engelbart of Stanford patents the mouse; he also develops the word processor.
1968	UNIX is developed at Bell Laboratories by Ken Thompson and Dennis Ritchie, ushering in the era of computing languages that aren't tied to a single computer brand.
1969	ARPANET, a four-computer network, is developed by the US Department of Defense, and forms the basis of what is now the Internet.
1971	Intel introduces the 4004 chip, the first microprocessor.
1972	Bill Gates drops out of Harvard. With Paul Allen he starts Traf-O-Data, which makes traffic-counting systems. The company later becomes Microsoft Corp. Ray Tomlinson modifies an e-mail program he created in 1971 for ARPANET where it becomes a big hit. The @ sign was chosen from the punctuation keys on Tomlinson's Model 33 teletype for its 'at' meaning. Atari releases 'Pong', the first video game.
1973	Vinton Cerf and Bob Kahn present the basic Internet ideas at the University of Sussex.
1974	The Altair, the first popular personal computer, is developed by Edward Roberts of Micro Instrumentation Telemetry Systems and sold as a kit.

1975	IBM introduces its Mercury 5100 portable computer. The machine costs $9000 and weighs 25 kilograms.
1976	Wang unveils the first fully computerized word-processing machine. Price tag: $30 000.
	Steve Jobs and Steve Wozniak found Apple Computer Inc., working out of Jobs's garage.
1979	Dan Bricklin and Bob Frankston develop VisiCalc, the first spreadsheet software. They later sell the rights to Lotus Development Corp.
	CompuServe starts a bulletin-board service for computer enthusiasts.
1980	IBM asks Microsoft for an operating system for its upcoming PC. Gates and Allen don't have one, so they license the rights to Q-DOS (which stands for Quick and Dirty Operating System) from Timothy Patterson for less than $100 000. Microsoft drops the Q and changes the name to Disk Operating System (DOS).
1981	IBM rolls out the IBM PC. The first graphic user interface, Star, is developed at Xerox Palo Alto Research Centre.
1983	Novell introduces NetWare, its network operating system.
1984	Hewlett Packard introduces the Laser Jet printer.
	The break-up of AT&T in the US gives way to competition and ultimately innovation in the world-wide telecom market.
	Apple unveils the Macintosh. The company runs just one ad, but news and talk shows replay it.
1985	Microsoft introduces Windows. IBM isn't interested, so Microsoft sells the system independently. Windows doesn't become popular until the release of version 3.0 in 1990.
	First addresses ending in .com launched.
1988	The US National Science Foundation Network, in collaboration with IBM, MCI and the University of Michigan, develops NFSnet. The system boosts the Internet's capacity and allows the transfer of graphics.
1989	The World Wide Web comes to be, developed by English computer scientist Timothy Berners Lee for international researchers at the European

	Organisation for Nuclear Research in Geneva. WWW software is released in 1992.
1991	The first text-only Web browser is released to the public. The Internet backbone, the fastest part of the whole system, runs at 44 Mbps.
1992	The term 'surfing the net' is coined by the writer Jean Armour Polly.
1993	Intel releases the Pentium processor. The first web browser, Mosaic, is released by the National Centre for Supercomputing Applications at the University of Illinois in Champaign.
1994	The Web is growing at an annual rate of 341 634%. Two Arizona-based lawyers, Lawrence Canter and Martha Siegel, send an advert to 6000 newsgroups – the first spam.
1995	Microsoft launches Windows95. Lycos and Altavista are launched.
1997	Deep Blue, an IBM supercomputer, beats world chess champion Garry Kasparov.
1998	A US court bans the practice of buying domain names of famous companies and then selling them for exorbitant prices.
1999	Microsoft used its monopoly in PC operating systems to hurt consumers and competitors, says a US judge's findings of fact.
2000	There are 21.1 million websites worldwide.

Sources: Collier's Encyclopaedia, Computer Chronicles, Microsoft Encarta, World Almanac

4

Valuation techniques for traditional common stocks

> . . . it used to be the rule of thumb that you shouldn't have a higher P/E on a growth stock than its growth rate . . . Ultimately 12%–15% growth doesn't hold up 50% multiples . . . I don't understand how you get a 23 market P/E with no growth in profits. (Scott Black, Barron's, 6 June 1998)

In order to gain a feel for the process of valuing Internet and technology stocks it is essential to start with how one values what could be referred to as traditional stocks.When analysing stock markets the starting point is not to confuse price with value. Price is what you pay: value is what you get. Investors do not (or should not) buy financial assets for emotional reasons – they buy them for profit. Value itself is directly related to the level and expected growth of those cash flows. The founding father of investment analysis, Ben Graham, in writing *The Intelligent Investor* (1949), neatly summed up the valuation dilemma: 'In the short term, the stock market is a voting machine: in the long term it's a weighing machine.'

While day-to-day market prices are largely driven by investors' emotions, long term they are based on the fundamental values of profits, earnings and cash flows.

Most analysts rely on the fundamental building blocks of stock value – the dividends and earnings of firms – to make stock valuation decisions. Stocks have value because of the potential cash flows, called dividends, which a stockholder expects to receive from ownership of the firm. Stocks also have value if, in the future, other stockholders may decide that the valuation of these future dividends is not fully reflected in the future share price. It is by forecasting and valuing potential future dividends and earnings and deciding whether someone will, in the future, value these differently, that enables one to judge the investment value of shares.

The uncertainty of future cash flows makes the pricing of stocks according to the system of applying net present values more difficult. In this chapter we examine the valuation principles applicable to common stocks. Two basic approaches of stock valuation when using *security analysis* are:

(1) *the present value approach*, or *capitalization of income method*; and
(2) *fundamental analysis*.

Present value analysis: capitalization of income method

The classic method of calculating intrinsic value applies present value analysis. This technique is often referred to as the capitalization of income method. The present value process involves the capitalization (discounting) of future cash flows. That is, the *intrinsic value* of a stock is equal to the present value of the future stream of cash flows an investor expects to receive from the asset. These cash flows could be periodic, such as dividends, or simply a terminal price or redemption value, or a combination of these. Since these cash flows occur in the future, they must be discounted at an appropriate rate to determine their present value. The sum of these discounted cash flows is the estimated intrinsic value of the stock. This is illustrated with Equation 4.1.

$$\text{Value}_{t=0} = \sum_{t=1}^{n} \frac{\textit{Cash flows}}{(1+k)^t} \qquad \text{Eq 4.1}$$

where

$\text{Value}_{t=0}$ = the value of the asset now (time period 0)

Cash flows = the future cash flows resulting from ownership of the stock

k = the appropriate discount rate of return required by an investor for an investment of this type

n = number of periods over which the cash flows are expected

To derive the intrinsic value using this model an investor must:

(1) Estimate an appropriate required rate of return.
(2) Estimate the amount and timing of the future stream of cash flows.
(3) Use these two components in a present value model to estimate the value of the stock. This is then compared to the current market price of the stock.

Figure 4.1 summarizes the present value process used. This emphasizes the factors that go into valuing common stocks. The exact present value process used by investors in the market place depends on assumptions made about the growth rate in the expected stream of cash flows, as explained later in this chapter.

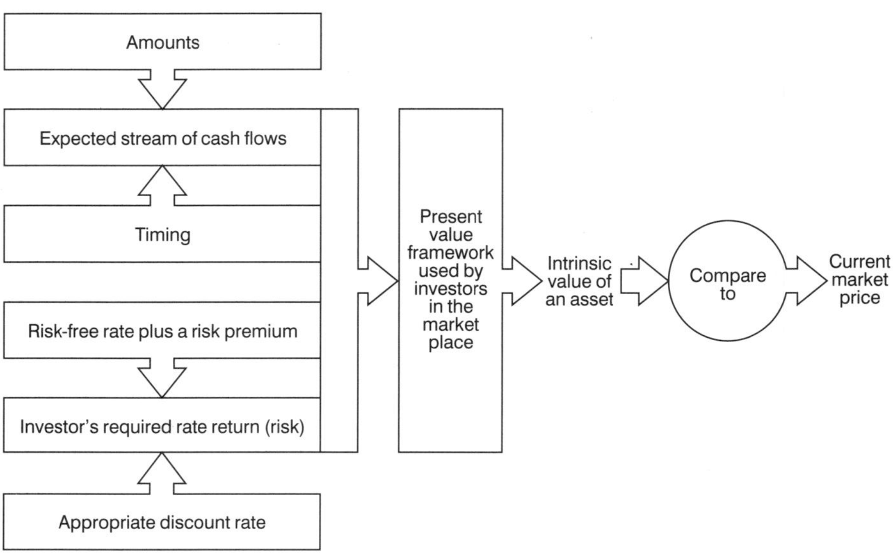

Figure 4.1 The present value approach to valuation

The required rate of return

An investor who is considering the purchase of a common stock must assess its risk and, given its risk, the *minimum expected rate of return* that will be required to induce the investor to make the purchase. This minimum expected return, or *required rate of return,* is an opportunity cost.

The required rate of return, capitalization rate and discount rate are interchangeable terms in valuation analysis. While in theory we know what this variable is, in practice it is not easy to determine the precise number to use. Because of this complexity, we will generally assume that we know the capitalization rate and concentrate here on the other issues involved in the valuation of stocks.

The expected cash flows

The other component that goes into the present value framework is the expected stream of cash flows. The value of common stock is the present value of all the cash flows to be received from the issuer. The questions that then arise are as follows:

(1) What are the cash flows to use in valuing a stock?
(2) What are the expected amounts of the cash flows?
(3) When will the expected cash flows be received?

Stockholders may plan to sell their shares at some time in the future, resulting in a cash flow from the sales price. As shown later, however, even if investors think of the total cash flows from common stocks as a combination of dividends and a future price at which the stock can be sold, this is equivalent to using the stream of all dividends to be received on the stock as the key valuation principle.

What about earnings? Are they important? Can they be used in valuing a stock? The answer to both questions is a clear Yes. Dividends are paid out of earnings, so earnings are clearly important. The second stock valuation approach, fundamental analysis, to be considered later, uses earnings and a P/E ratio to determine intrinsic value. Therefore, earnings are an important part of fundamental analysis: in fact, earnings receive more attention from investors than any other single variable.

If all earnings are paid out as dividends, they will be accounted for as dividends. If the company retains earnings they presumably will be reinvested, thereby enhancing future earnings and, ultimately, dividends. The present value analysis should not count the earnings reinvested currently and also count them again paid later as dividends. If they are properly defined and separated, these two variables produce the same results. This means that more than one present value model is possible. However, it is always correct to use dividends in the present value analysis, and this is what is almost always done when investors use the present value approach to stock valuation.

Because dividends are the only cash flow stream to be received directly by investors under normal conditions, it is appropriate to have a stock valuation model based on dividends. It is now important to consider such a model, the dividend discount model, which is the basis for understanding the fundamental valuation of common stocks.

The dividend discount model (DDM)

Since dividends are the only cash payment a stockholder receives directly from a firm, *they provide the foundation of valuation for common stocks.* Applying Equation 4.1 specifically to value common stocks, the cash flows are the dividends *expected* to be paid in each future period. An investor or analyst using this approach carefully studies the future prospects for a company and estimates the likely dividends to be paid. In addition, the analyst estimates an appropriate required rate of return or discount rate based on the risk foreseen in the dividends, taking account of the alternatives available. Finally, the investor would discount to the present the entire stream of estimated future dividends, properly identified as to amount and timing.

Equation 4.1 adapted for common stocks where dividends are the cash flows results in Equation 4.2. This equation, known as the dividend discount model (DDM), states that the value of a stock today is the discounted value of all future dividends:

$$\hat{P}_{cs} = \frac{D_1}{(1 + k_{cs})} + \frac{D_2}{(1 + k_{cs})^2} + \frac{D_3}{(1 + k_{cs})^3} + \frac{D_\alpha}{(1 + k_{cs})^\infty}$$

$$= \sum_{t=1}^{n} \frac{D_t}{(1 + k_{cs})^t} \qquad \text{Eq 4.2}$$

$$= \textit{dividend discount model}$$

where

$\hat{P}_{cs}$ = intrinsic value or estimated value of a common stock today based on the model user's estimates of the future dividends and the required rate of return

$D_1, D_2, \ldots$ = the dividends expected to be received in each future period

k_{cs} = the required rate of return for this stock, which is the discount rate applicable for an investment with this degree of riskiness

There are two immediate problems with Equation 4.2:

(1) The last term in Equation 4.2 indicates that investors are dealing with infinity. They must value a stream of dividends that may be paid forever, since common stock has no maturity date.
(2) The dividend stream is uncertain:
  (a) There are no specified number of dividends, if in fact any are paid at all. Dividends must be declared periodically by the firm's board of directors.
  (b) The dividends for most firms are expected to grow over time; therefore, investors usually cannot simplify Equation 4.2 to a perpetuity. Only if dividends are not expected to grow can such a simplification be made.

How are these problems resolved? The first problem, that Equation 4.2 involves an infinite number of periods and dividends, will be resolved when we deal with the second problem, specifying the expected stream of dividends. However, from a practical standpoint this problem is not as troublesome as it appears. At reasonably high discount rates, such as 12%, 14%, or 16%, dividends received 40 or 50 years in the future are worth very little today, so that investors need not worry about them. For example, the present value of $1 to be received 50 years from now, if the discount rate is 15%, is 0.0009 cents.

The conventional solution to the second problem, that the dollar amount of the dividend is expected to grow over time, is to make some assumptions about the *expected growth rate* of dividends. That is, the investor or analyst estimates or models the expected percentage rate of growth in the future stream of dividends. To do this, they classify each stock to be valued into one of three categories based on the expected growth rate in dividends. In summary: the dividend discount model is operationalized by estimating the expected growth rate(s) in the dividend stream.

All stocks that pay a dividend, or that are expected to pay dividends some time in the future, can be modelled using this approach. It is critical to remember in using the DDM that an investor must account for all dividends from now to infinity by modelling the growth rate(s). As shown below, the mechanics of this process are such that we don't actually see all of these dividends because the formulas reduce to a simplified form, but nevertheless we are accounting for all future dividends when we use the DDM.

It is necessary in using the DDM to remember that the dividend currently being paid on a stock (or the most recent dividend paid) is designated as D_o and is, of course, known. Specifically, D_o, designates the current dividend being paid, or the most recent dividend paid. Investors must estimate the future dividends to be paid, starting with D_1 the dividend expected to be paid in the next period.

The three growth rate models for dividends are:

(1) **The zero growth rate model**: a dividend stream with a zero growth rate resulting from a fixed dollar dividend equal to the current dividend being paid, D_0, being paid every year from now to infinity. This is typically referred to as the no growth rate or zero growth rate model:

D_0 D_0 D_0 D_0 + · · · + D_0	*Dividend stream*
0 1 2 3 + · · · + ∞	*Time period*

(2) **The constant growth rate model**: under this model we have a dividend stream that is growing at a constant rate g, starting with D_0. This is typically referred to as the constant or normal growth version of the dividend discount model:

D_0	$D_0(1+g)^1$	$D_0(1+g)^2$	$D_0(1+g)^3$	+ · · · + $D_0(1+g)^\infty$	*Dividend stream*
0	1	2	3	+ · · · + ∞	*Time period*

(3) **The multiple growth rate model**: under this model we have a dividend stream that is growing at variable rates, for example, g_1, for the first four years and g_2 thereafter. This is referred to as the multiple growth version of the dividend discount model:

D_0	$D_1 = D_0(1+g_1)$	$D_2 = D_1(1+g_1)$	$D_3 = D_2(1+g_1)$	$D_4 = D_3(1+g_1)$
0	1	2	3	4

$D_5 = D_4\,(1 + g_2) + \cdots + D_\infty = D_{\infty-1}\,(1 + g_2)$	*Dividend stream*
$5 \qquad + \cdots + \qquad \infty$	*Time period*

The zero growth rate model

The *fixed dollar* dividend model reduces to a perpetuity. Assuming a constant dollar dividend, Equation 4.2 simplifies to the *no growth model* shown as Equation 4.3.

$$\hat{P}_0 = \frac{D_0}{k_{cs}} = \begin{array}{c}\textit{Zero growth version} \\ \textit{of the dividend discount model}\end{array} \qquad \text{Eq 4.3}$$

where D_0 is the constant dollar dividend expected for all future time periods and k_{cs} is the opportunity cost or required rate of return for this particular common stock and $\hat{P}_0$ is the current price.

A zero growth rate common stock is a perpetuity and is easily valued once k_{cs} is determined.

It is extremely important in understanding the valuation of common stocks using the DDM to recognize that in all cases an investor is discounting the future stream of dividends from now to infinity. This fact tends to be overlooked when using the perpetuity formula involved with the zero growth rate case because the discounting process is not visible. Nevertheless, we are accounting for all dividends from now to infinity in this case, as in all other cases. It is simply a mathematical fact that dividing a constant dollar amount by the discount rate, k_{cs}, produces a result equivalent to discounting each dividend from now to infinity separately and summing all of the present values.

The constant growth rate model

The other two versions of the DDM, the constant growth model and the multiple growth case, demonstrate that to establish the cash flow stream of expected dividends, which is to be subsequently discounted, it is first necessary to compound some beginning dividend into the future. Obviously, the higher the growth rate used, the greater the future amount. Furthermore, the longer the time period, the greater the future amount.

A well-known scenario in valuation is the case in which dividends are expected to grow at a constant rate over time. This *constant or normal growth model* is shown as Equation 4.4

$$\hat{P}_0 = \frac{D_0\,(1+g)}{(1+k_{cs})} + \frac{D_0\,(1+g)^2}{(1+k_{cs})^2} + \frac{D_0\,(1+g)^3}{(1+k_{cs})^3} + \cdots + \frac{D_0\,(1+g)^\infty}{(1+k_{cs})^\infty} \quad \text{Eq 4.4}$$

where D_0 is the current dividend being paid and growing at the constant rate g, and k_{cs} is the appropriate discount rate.

Equation 4.4 can be simplified to the following equation:

$$\hat{P}_0 = \frac{D_1}{k-g} = \frac{\textit{Constant growth version of}}{\textit{the dividend discount model}} \quad \text{Eq 4.5}$$

where D_1 is the dividend expected to be received at the end of Year 1.

Equation 4.5 is used whenever the *growth rate* of future dividends is estimated to be a constant. In actual practice, it is used quite often because of its simplicity and because it is the best description of the actual behaviour of a large number of companies and, in many instances, the market as a whole. Example 4.1 illustrates how the constant growth model can be applied to value stocks.

Example 4.1

Assume Nadia Corporation is currently paying \$1 per stock in dividends and investors expect dividends to grow at the rate of 7% a year for the foreseeable future. For investments at this risk level, investors require a return of 15% a year. The estimated price of Nadia Corporation is:

$$\hat{P}_0 = \frac{D_1}{k-g}$$

$$= \frac{\$1.00\ (1.07)}{0.15 - 0.07} = \$14.38$$

Note that a current dividend (D_0) must be compounded one period because the constant growth version of the DDM specifies the numerator as the dividend expected to be received one period from now, which is (D_1). In valuation terminology, D_0 represents the dividend currently being paid, and D_1 represents the dividend expected to be paid in the next period. If D_0 is known, D_1 can always be determined:

D_0 = Current dividend $D_1 = D_0(l + g)$

where g is the expected growth rate of dividends.

Table 4.1 Present value of 60 years of dividends (current dividend = \$1, $g = 7\%$, $k = 15\%$)

Period	*Dollar dividend*	*PV factor*	*PV of dollar dividend*
1	1.07	0.8696	0.93
2	1.14	0.7561	0.87
3	1.23	0.6576	0.81
4	1.31	0.5718	0.75
5	1.40	0.4972	0.70
6	1.50	0.4323	0.65
7	1.61	0.3759	0.60
8	1.72	0.3269	0.56
9	1.84	0.2843	0.52
10	1.97	0.2472	0.49
11	2.10	0.2149	0.45
12	2.25	0.1869	0.42
13	2.41	0.1625	0.39
14	2.58	0.1413	0.36
15	2.76	0.1229	0.34
16	2.95	0.1069	0.32
17	3.16	0.9385	0.29
18	3.38	0.0808	0.27
19	3.62	0.0703	0.25
20	3.87	0.0611	0.24
21	4.14	0.0531	0.22
22	4.43	0.0462	0.20
23	4.74	0.0402	0.19
24	5.07	0.0349	0.18
25	5.43	0.0304	0.16
26	5.81	0.0264	0.15
27	6.21	0.0230	0.14
28	6.65	0.0200	*0.13*
29	7.11	0.0174	0.12
30	7.61	*0.0151*	0.11
31	8.15	0.0131	0.11
32	8.72	0.0114	0.10
33	9.33	0.0099	0.09
34	9.98	0.0086	0.09
35	10.68	0.0075	0.08
36	11.42	0.0065	0.07
37	12.22	0.0057	0.07
38	13.08	0.0049	0.06
39	13.99	0.0043	0.06
40	14.97	0.0037	0.06
41	16.02	0.0032	*0.05*
42	17.14	0.0028	0.05
43	18.34	0.0025	0.*05*
44	19.63	0.0021	0.04
45	21.00	0.0019	0.04
46	22.47	0.0016	0.04
47	24.05	0.0014	0.03
48	24.73	0.0012	0.03
49	27.53	0.0011	0.03
50	29.46	0.0009	0.03
51	31.52	0.0008	0.03
52	33.73	0.0007	0.02
53	36.09	0.0006	0.02
54	38.61	0.0005	0.02
55	41.32	0.0005	0.02
56	44.21	0.0004	0.02
57	47.30	0.0003	0.02
58	50.61	0.0003	0.02
59	54.16	0.0003	0.01
60	57.95	0.0002	0.01
Sum of dividends	\$870.47		\$13.20

Sum of first 60 years of discounted dividends is \$13.20

To completely understand the constant growth model, which is widely used in valuation analysis, it is instructive to think about the process that occurs under constant growth. Table 4.1 illustrates the case of Nadia's growth stock with a current dividend of $1 per share ($D_0$), an expected constant growth rate of 7%, and a required rate of return (k) of 15%. The present value factor is taken from the standard statistical tables.

As Table 4.1 shows, the expected dollar dividend for each period in the future grows by 7%. Therefore, D_1 = $1.07, D_2 = $1.14, D_3 = $1.23, and so forth. Only the first 60 years of growth are shown, at the end of which time the dollar dividend is $57.95. The last column of Table 4.1 shows the discounted value of each of the first 60 years of dividends. Thus, the present value of the dividend for Period 1, discounted at 15%, is $0.93, while the present value of the actual dollar dividend in Year 60 of $57.95, is worth only $0.01in today's money.

The message from Table 4.1 is that dividends received far in the future, assuming normal discount rates, are worth very little today. It is useful to check Table 4.1 in detail. It provides a simple technique for understanding the principles of stock valuation.

The estimated price of Nadia Corporation as illustrated in Table 4.1 is the sum of the present values of each of the future dividends. Adding each of these present values together from now to infinity would produce the correct estimated value of the stock. Note from Table 4.1 that adding the present values of only the first 60 years of dividends together produces an estimated value of only $13.20. The correct answer, as obtained from adding all years from now to infinity, and applying Equation 4.6, is:

$$\hat{P}_0 = \textit{Estimated price} = \frac{\$1.07}{0.15 - 0.07} = \$13.38 \qquad \text{Eq 4.6}$$

As can be seen, the years beyond 40 to 50 typically add very little to the estimated value of a stock. Adding all of the discounted dividends together for the first 60 years produces a present value, or estimated value for the stock, of $13.20, which is only $0.18 different from using Equation 4.6. To repeat, years 61 to infinity add a total value of only $0.18 to the stock price.

Example 4.1 illustrates a very important point about these valuation models that was explained earlier. The constant growth version takes account of all future cash flows from now to infinity, although this is not apparent from simply looking at the equation itself. Although Equation 4.6 has no summation or infinity sign, the results produced by this equation are equivalent to those that would be obtained if the dividend for each future period is determined and then discounted back to the present. Again, the mathematics of the process masks the fact that all dividends from now to infinity are being accounted for.

To fully understand the constant growth rate version of the DDM, it is also important to realize that the model implies that the stock price for any one period is estimated to grow at the same rate as the dividends, which is g. This means that the growth rate in price plus the growth rate in dividends will equal k, the required rate of return.

For Nadia Corporation the estimated price today is \$13.38 and for the end of Period 1, using D_2 in the numerator of Equation 4.6 gives us:

$$\hat{P}_0 = \frac{(\$1.07)\ (1.07)}{0.15 - 0.07}$$

$$= \$14.3$$

This estimated price at the end of Period 1 is 7% higher than the estimated price today of \$13.38.

$$\textit{Price change} = \frac{\textit{Ending price} - \textit{Beginning price}}{\textit{Beginning price}}$$

$$= (\$14.31 - \$13.38)\ /\ \$13.38 = 7\%$$

An examination of Equation 4.6 quickly demonstrates the factors affecting the price of a common stock, assuming that we are assuming the constant growth version of the DDM to be the applicable valuation approach. The lessons of this version are:

1 If the market lowers the required rate of return for a stock, the stock price will rise (other things being equal).
2 If investors decide that the expected growth in dividends will be higher as the result of some favourable development for the firm, the stock price will also rise (other things being equal).

Of course, the converse for these two situations also holds – a rise in the discount rate or a reduction in the expected growth rate of dividends will lower stock prices.

The present value or intrinsic value calculated from Equation 4.6 is quite sensitive to the estimates used by the investor in the equation. Relatively small variations in the inputs can change the estimated price by large percentage amounts. This is illustrated in Examples 4.2, 4.3 and 4.4.

Example 4.2

For Nadia Corporation assume the discount rate used, *k*, is 16% instead of 15%, with the other variables held constant:

$$\frac{\$1(1.07)}{0.16 - 0.07}$$

$$P = \$11.89$$

In this example, a 1 percentage point rise in *k* results in an 11.14% decrease in price, from today's price of $13.38 to $11.89.

Example 4.3

Assume that for Nadia Corporation the growth rate, *g*, is 6% instead of 7%, with the other variables held constant:

$$\frac{\$1(1.06)}{0.15 - 0.06} = \$11.77$$

In this example, a 1 percentage point decline in *g* results in a 12% decrease in price, from $13.38 to $11.77.

Example 4.4

Assume that for Nadia Corporation the discount rate rises to 16% and the growth rate declines to 4%:

$$\frac{\$1(1.04)}{0.16 - 0.04} = \$8.67$$

In this example, the price declines from today's price of $13.38 to $8.67, a 35% change.

These wide differences suggest why stock prices constantly fluctuate as investors make their buy and sell decisions. Even if all investors use the constant growth version of the DDM to value a particular common stock, many different estimates of value will be obtained given that:

(1) Each investor has his or her own required rate of return, resulting in a relatively wide range of values of k.
(2) Each investor has his or her own estimate of the expected growth rate in dividends. Although this range may be reasonably narrow in most valuation situations, small differences in g can produce significant differences in price, everything else remaining constant.

Thus, at any point in time for a particular stock, some investors are willing to buy, whereas others wish to sell, depending on their evaluation of the stock's prospects. This helps to make markets active and liquid.

The multiple growth rate model

Many firms grow at a rapid rate (or rates) for a number of years and then slow down to an 'average' growth rate. This was certainly the case for technology stocks. Other companies pay no dividends for a period of years, often during their early growth period. The constant growth model, discussed earlier, is unable to deal with these situations; therefore, a different model is needed that can handle these possible scenarios. Such a variation of the DDM is the multiple growth rate model.

Multiple growth is defined as a situation in which the expected future growth in dividends must be described using two or more growth rates. Although any number of growth rates is possible, most stocks can be described using two or possibly three. It is important to remember that we need to assume at least two different growth paths. This is the distinguishing characteristic of multiple growth situations.

A number of companies have experienced rapid growth that could not be sustained forever. During part of their lives their growth exceeded that of the average company in the economy, but later the growth rate slowed. Examples from the past would include McDonald's, Disney, IBM, Microsoft and Amazon.

To capture the expected growth in dividends under this scenario it is necessary to model the dividend stream during each period of differential growth. It is reasonable to assume that at some point the company's growth will slow down to that of the economy as a whole. At this time the company's growth can be described by the constant growth model. What remains, therefore, is to model the exact dividend stream up to the point at which dividends slow to a normal growth rate and then to find the present value of all the components.

A well-known multiple growth model is the two-stage growth rate model. This model assumes near-term growth at a rapid rate for some period (typically, 2 to 10 years) followed by a steady long-term growth rate that is sustainable (i.e., a constant growth rate as discussed earlier). Equation 4.7 describes the multiple growth model:

$$\hat{P}_0 = \sum_{t=1}^{n} \frac{D_0 (1 + g_1)}{(1 + k)^t} + \frac{D_n (1 + g_c)}{k - g_c} \frac{1}{(1 + k)^n} \qquad \text{Eq 4.7}$$

where

$\hat{P}_0$ = the intrinsic or estimated value of the stock today
D_0 = the current dividend
g_1 = the supernormal (or subnormal) growth rate for dividends
g_c = the constant growth rate for dividends
k = required rate of return
n = the number of periods of supernormal (or subnormal) growth
D_n = the dividend at the end of the abnormal growth period

Notice in Equation 4.7 that the first term on the right-hand side defines a dividend stream covering n periods, growing at a high (or low) growth rate of g, and discounted at the required rate of return, k. This term covers the period of supernormal growth, at which time the dividend is expected to grow at a constant rate forever. The second term on the right-hand side is the constant growth version discussed earlier, which takes the dividend expected for the next period, $n + 1$, and divides it by the difference between k and g. Notice, however, that the value obtained from this calculation is the value of the stock at the beginning of period $n + 1$ (or the end of period n), and it must be discounted back to time period zero by multiplying by the appropriate discount (present value) factor. Conceptually, the valuation process being illustrated here is:

$\hat{P}_0$ = discounted value of all dividends through the unusual growth period n plus the discounted value of the constant growth model which covers the period $n + 1$.

It is useful to think about the second term in Equation 4.7 as representing a $\hat{P}_0$ or the expected price of the stock derived from the constant growth model as of the end of period n. The constant growth version of the dividend discount model is used to solve for expected price at the end of period n, which is the beginning of period $n + 1$. Therefore,

$$P_n = \frac{D_{n+1}}{k - g_c}$$

Because P is the expected price of the stock at the end of period n, it must be discounted back to the present. When added to the value of the discounted dividends from the first term, the estimated value of the stock today (P_0) is produced. Example 4.5 illustrates the multiple growth model.

Example 4.5

Figure 4.2 illustrates the concept of valuing a multiple-growth rate company. In this example, the current dividend is \$1 and it is expected to grow at the higher rate (g_1) of 12% a year for 5 years, at the end of which time the new growth rate (g_c) is expected to be a constant 6% a year. Assume the required rate of return is 10%.

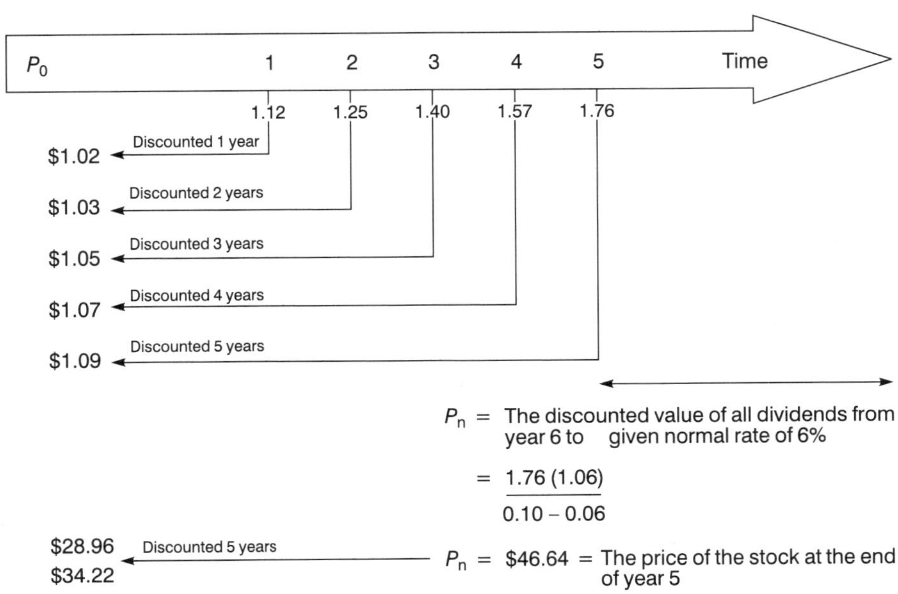

Figure 4.2 Valuing a multiple growth rate company

The first step in the valuation process, illustrated in Figure 4.2, is to determine the dollar dividends in each year of supernormal growth. This is done by compounding the beginning dividend, $1, at 12% for each of five years, producing the following expected dividend stream:

D_0 = $1.00

$$
\begin{aligned}
D_0 &= \$1.00(1.12) &= \$1.12 \\
D_1 &= \$1.00(1.12)^2 &= \$1.25 \\
D_2 &= \$1.00(1.12)^3 &= \$1.40 \\
D_3 &= \$1.00(1.12)^4 &= \$1.57 \\
D_4 &= \$1.00(1.12)^5 &= \$1.76
\end{aligned}
$$

Once the stream of dividends over the supergrowth period has been determined, they must be discounted to the present using the required rate of return of 10%. Thus

$$
\begin{aligned}
\$1.12(0.909) &= \$1.02 \\
\$1.25(0.826) &= \$1.03 \\
\$1.40(0.751) &= \$1.05 \\
\$1.57(0.683) &= \$1.07 \\
\$1.76(0.621) &= \underline{\$1.09} \\
 & \$5.26
\end{aligned}
$$

Summing the five discounted dividends produces the value of the stock for its first five years only, which is $5.26. To evaluate Years 6 and later, when constant growth is expected, the constant growth model is used.

$$
\begin{aligned}
P_n &= \frac{D_{n+1}}{k - g_c} \\
&= \frac{D_6}{k - g_c} \\
&= \frac{D_5\ (1.06)}{k - g_c} \\
&= \frac{1.76\ (1.06)}{0.10 - 0.06} \\
&= \$46.64
\end{aligned}
$$

Thus, $46.64 is the expected price of the stock at the beginning of Year 6 (end of Year 5). It must be discounted back to the present, using the present value factor for five years and 10%, which is 0.621,and can be found from the standard statistical tables. Therefore,

$$P_n \text{ discounted to today} = P_n \text{ (PV factor for 5 years, 10\%)}$$

$$= \$46.64\ (0.621)$$

$$= \$28.96$$

The last step is to add the two present values together:

\$5.26 = present value of the first five years of dividends

\$28.96 = present value of the price at the end of Year 5, representing the discounted value of dividends from Year 6 to ∞

= \$34.22 = $\hat{P}_0$, the present value of this multiple growth rate stock

Fundamental analysis for valuing stocks

We have thus far discussed the process of applying valuation models to stock valuation – an approach often referred to as the capitalization of income method. The remainder of this chapter is devoted to other valuation techniques, often referred to collectively as fundamental analysis for stock valuation.

As already discussed, the value of any asset is determined by the discounted value of all future expected cash flows. Discounted means that future cash flows are not valued as highly as current flows. For stocks, cash flows come primarily in the form of dividends, but occasionally from other distributions resulting from the sale of assets or other transactions. For most assets – and especially for all stocks – the future cash flows are uncertain and depend on the financial circumstances of the firm.

As discussed above, the future cash flows from assets are discounted because cash received in the future is not worth as much as cash received in the present. There are many reasons for this: the innate preferences of most individuals to enjoy their consumption now rather than wait for tomorrow; productivity, which allows funds invested today to yield a higher return tomorrow; inflation, which reduces the future purchasing power of cash received in the future; and the uncertainty associated with any event which takes place in the future. The rate at which future cash flows from stocks are discounted is always referred to as the discount rate for stocks.

Since dividends generally increase over time, the value of most stocks depends on what may happen many decades hence.

Estimates of *all* future cash flows are important for the valuation of stocks – not only the ones received during the time the investor holds the asset. However, for a short-term investor, the return on an investment will depend not only on their assessment of the cash flows but also on the market's assessment of the cash flows at the time of the sale. This is because a large part of the return for a short-term stockholder comes from the proceeds of the sale of the stock and not from the dividends received. Unless one intends to hold the stock forever, one must take into account how much other investors in the market will value the stock at the time of its sale in order to estimate one's return.

As a consequence, for most investors, the return on a stock is composed of two parts: the cash flows received while holding the stock and the change in valuation of the stock by the market from the time of purchase to the time of sale. An investor might be enthusiastic about the dividend prospects of a certain stock, but if he expects to sell the stock next year, he is at the mercy of what *other* investors will think of the firm's prospects at that time.

Most investors attempt to profit in the market by buying a stock with what they consider attractive future returns, hoping that other investors will come to agree with their judgement. If this comes to pass, the price of the stock will rise. In fact, the fastest way to make money in the market is to successfully forecast how *other* investors might change the basis on which *they* make judgements about stocks' values in the near future. For example, if you think that other investors will turn optimistic about technology stocks, then you may make money buying such stocks whether or not you believe they are good value.

Accordingly, success for short-term investors comes primarily from discerning how the public will view stocks in the future, and quite secondarily from the cash flows realized by the investment itself. The tension between investing in the long run and the short run was best described by John Maynard Keynes when he wrote in 1936:

> most of these [professional investors and speculators] are, in fact, largely concerned, not with making superior long-term forecasts of the probable yield of an investment over its whole life, but with foreseeing changes in the conventional basis of valuation a short time ahead of the public. They are concerned, not with what an investment is really worth to a man who buys it 'for keeps', but with what the market will value it at, under the influence of mass psychology, three months or a year hence.

The game of forecasting future investor sentiment is difficult and deters many from investing in stocks. But, you need not forecast market sentiment in order to profit by investing in stocks. Although investment advice geared to the short run hinges on predicting the judgement of other investors, in the long run one can ride out the waves of investor sentiment and profit from your own predictions. Investing for the long run does however necessitate, first, a knowledge of stock market history and secondly, the ability to apply some suitable stock valuation models.

So we need to know some stock market history, and given that background we must then ask whether there are any other stock valuation models available which would enable us to compare our results with those of the capitalization of income method already discussed. So does history tell us anything about the long run performance of US stock markets?

What do we know about US stock market history?

When predicting future returns on the market, the price-to-earnings (P/E) ratio and the dividend yield, compared to historical standards, are the most frequently cited criteria for judging the valuation of the stock market.

Table 4.2 displays the dividends and earnings yield on stocks represented by the US S&P 500 Index for the period 1871–1994. It can be seen that the average historical earnings yield during this period is 8.1%, giving an average price/earnings ratio of 13.6. This means that investors priced stocks on average at about 12 to 15 times their annual earnings. Despite the stability of the average, the P/E ratio for the period has shown considerable annual variation, ranging from a high of 27 in 1894 to a low of 6 in 1950. (It should be noted that P/E ratios have been around 30 towards the end of the last millennium and the beginning of the new one.)

Table 4.2 Historical averages of dividend yield and earnings yield, 1871–1994

Period	*Dividend yield (%)*	*Price/ earnings*	*Earnings yield (%)*	*Period*	*Dividend/ earnings (%)*	*Growth of real dividends (%)*	*Growth of real earnings (%)*	*Growth of real S&P (%)*
1871–1994	4.9%	13.6	8.1%	1871–1994	64.0%	1.2%	1.5%	1.8%
1871–1958	5.3%	13.4	8.2%	1871–1958	71.5%	0.7%	1.0%	0.8%
1959–1994	3.8%	14.1	7.8%	1959–1994	51.8%	2.0%	2.1%	3.4%

Source: Siegel, 1995

Another yardstick used to judge the future of the market is the dividend yield. The dividend yield has averaged 4.9% from 1871, with a high of 8.71% in 1959 and a low of 2.85% in 1973. The dividend yield shows far less variability than the earnings yield, as managers pursue policies to stabilize the cash payments to stockholders.

In contrast to the long-term stability of the earning yield, the US dividend yield has fallen significantly in recent years. Between 1959 and 1994, the average dividend yield was 3.8%, and has fallen since to below 2%. The primary cause of the reduction in the dividend yield is not the higher valuation of shares, but the reduction in the dividend pay out ratio – the fraction of earnings firms pay as dividends. The increase in US personal and corporate taxes has made it advantageous for firms to retain earnings, rather than pay them to stockholders in the form of taxable dividends.

The fundamental valuation techniques available

Numerous fundamental stock valuation methods are available for investors to choose between. It must be said that stock picking, as it is known, is an art rather than a science. However there are scientific elements to the exercise and this section summarizes the techniques available. It must be stressed that each technique has limitations and these should be kept in mind.

The fundamental techniques discussed below cover:

- The price/earnings ratio
- The price/earnings growth factor (PEG ratio)
- The dividend yield
- The yield ratio: bond yield to the dividend yield
- Tobin's q ratio
- Price to book value
- Price to sales ratio

The price earnings (P/E) ratio

The cash flows which are received by stockholders come primarily from dividends, and these dividends in turn come from earnings. Obviously, then, two of the most important financial statistics about stocks are earnings and dividends. The earnings per share divided by the current stock price is referred to as the earnings yield on the stock. Many investors prefer to think in terms of the inverse of the earnings yield – called the price/earnings, or P/E ratio. The P/E ratio, also called the earnings multiple, measures how many times current (or prospective) earnings the market is paying for a stock. The P/E ratio is the most popular fundamental variable for valuing both stocks and the whole market.

There are many measures of profit, but a particularly useful definition is earnings – the profit after taxes and any other changes that are attributable to ordinary shareholders. Earnings are reported by companies on a per share basis, and that can be translated by analysts into an earnings yield, to parallel the dividend yields, discussed below. As mentioned above, the ratio is presented the other way round as the price–earnings ratio. Thus a 5% earnings yield represents a P/E ratio of 20.

So the P/E ratio indicates how many years' purchase of a company's earnings are reflected in its share price. The higher the ratio the more optimistic investors must be about growth in order to pay the market price. Thus growth stocks, most recently represented by technology stocks, have been on high P/Es. Stocks on low P/Es may represent bargains, but more likely they are stocks facing troubled times.

Determinants of the P/E ratio

What determines the P/E ratio? The P/E ratio can be derived from the dividend discount model,discussed earlier.

Start with Equation 4.1, the estimated price of a stock using the constant growth version of the model. We use P_0 to represent estimated price from the model.

$$P_0 = \frac{D_1}{k - g}$$

Dividing both sides of Equation 4.1 by expected earnings, E_1, gives us Equation 4.8

$$P_0/E_1 = \frac{D_1/E_1}{k - g} \qquad \text{Eq 4.8}$$

Equation 4.8 indicates that those factors that affect the estimated P/E ratio are:

1 The dividend pay out ratio, D/E.
2 The required rate of return, k.
3 The expected growth rate of dividends, g.

The following relationships should then hold, other things being equal:

1 The higher the pay out ratio, the higher the P/E ratio.
2 The higher the expected growth rate, *g*, the higher the P/E ratio.
3 The higher the required rate of return, *k*, the lower the P/E ratio.

It is important to remember the phrase 'other things being equal' because usually other things are not equal and the preceding relationships do not hold by themselves. It is quite obvious, upon reflection, that if a firm could increase its estimated P/E ratio, and therefore its market price by simply raising its pay out ratio, it would be very tempted to do so. However, such an action would in all likelihood reduce future growth prospects, lowering *g*, and thereby offsetting the increase in the pay out ratio. Similarly, trying to increase *g* by taking on particularly risky investment projects would cause investors to demand a higher required rate of return, thereby raising *k*. Again this would work to offset the positive effects of the increase in *g*.

Factors 2 and 3 above are typically the most important factors in the determination of the P/E ratio because a small change in either can have a large effect on the P/E ratio.This is illustrated in Example 4.6.

Example 4.6

Assume that the pay out ratio is 60%. By varying *k* and *g*, and therefore changing the difference between the two (the denominator in Equation 4.8), investors can assess the effect on the P/E ratio as follows:

Assume $k = 0.15$ and $g = 0.07$

$$\text{P/E} = \frac{D_1/E_1}{k - g}$$

$$\text{P/E} = \frac{0.60}{0.15 - 0.07}$$

$$= \frac{0.60}{0.08} = 7.5$$

Now assume $k = 0.16$ and $g = 0.06$

$$\text{P/E} = \frac{0.60}{0.10} = 6$$

or that $k = 0.14$ and $g = 0.08$ $$P/E = \frac{0.60}{0.06} = 10$$

Small changes in our assumptions can be seen to imply large changes in the P/E ratio!

What are the limitations of P/E ratios?

There are several limitations to P/E ratios. They do not tell you:

- at what rate a company's earnings are expected to grow;
- whether a company's earnings growth is expected to accelerate or decelerate;
- how its earnings generating capability compares with other investments offering the same risk-reward relationship;
- exactly how earnings between companies differ. Earnings suffer from fundamental measurement problems because they depend on subjective judgement made by companies and their auditors. In bull markets accounting standards often deteriorate as companies bend the rules to meet the over-ambitious expectations of investors;
- how to adjust the ratio when accounting for massive structural change within an economy. As discussed in Chapters 5 and 6, the rising importance of intangible assets has meant that new methods of accounting need to be introduced. This forms a key component of the impact of the so-called New Economy.

Prospective price/earnings ratio

Some analysts calculate the prospective price/earnings ratio based on what they predict companies will earn in the future. Since expectations of future earnings differ between analysts it is not easy to apply this method for the overall stock index, but it may be useful for individual stocks and sectors.

Price/earnings growth factor (PEG)

A concept that has become popular in the past few years is the price/earnings growth factor, or PEG. This is the P/E ratio divided by the annualized growth rate of earnings.

Dividing the P/E ratio by the estimated growth of earnings is a gauge of how quickly the company is growing. This is known as the prospective PEG. For instance, two companies may have a P/E ratio of 20, but show different rates of earnings growth. Company A has earnings growth of 20%, while Company B has earnings growth of 5%. Company A's PEG ratio would be 1 (20 divided by 20), whereas Company B would have a PEG of 4 (20 divided by 5). The lower the PEG ratio, the better the potential value of the company. In this example, company A is growing very quickly and could represent better potential.

A PEG of significantly less than 1 is regarded as attractive, so a share with a moderate P/E ratio of 15 would need a growth rate of close to 20% a year to be a strong buy.

PEG can only be used for growth stocks. It would be inapplicable for cyclical stocks or turnaround situations. When calculating the estimated sustainable growth rate in earnings it is expected that these can be sustained for over 5 years. So a company with a P/E of 10, for example, growing at 20%, would have a PEG of 0.5 or half the estimated growth rate. A PEG of less than one, as discussed above, usually means that the stock is good value.

The essential drawback in valuation with all multiples is that they do not take account of changes in inflation and interest rates. Increases in interest rates will usually cause stock prices to fall, as the real net present value of companies' future earnings streams is eroded. Falling interest rates have the opposite effect.

Dividend yield

Table 4.3 illustrates that stocks earned significantly higher average returns between 1927 and 1995 than other financial assets such as corporate bonds or US Treasury Bills. While the stock returns shown in the Panel A of Table 4.3 are impressive, investors are more concerned with inflation-adjusted returns. Panel B of Table 4.3 presents average real returns after adjusting for inflation.

One approach to deciding whether the stock market is overvalued is to examine the dividend yield, defined as the rate of dividends to stock prices. When the dividend yield is low, stock prices are considered relatively high and vice versa.

The cash income from stocks is the most basic foundation for stock valuation. The dividends and cash flows from the stock are rarely identical to the earnings. Some earnings are usually retained by management to generate funds for future operations or expansion, or sometimes used to repurchase shares. The current cash return to the stockholder is called the

Table 4.3 Nominal and real returns (% per year) on US financial assets, 1927–95

	1927–95	*1946–95*
A: Nominal returns		
Stocks	9.54	11.23
Corporate bonds	5.39	5.46
Government bonds	4.85	5.04
Treasury bills	3.64	4.72
B: Real returns		
Stocks	6.43	7.03
Corporate bonds	2.28	1.25
Government bonds	1.74	0.84
Treasury bills	0.52	0.52

Source: Cochrane, 1997b

dividend yield and is represented by the dividend per share divided by the current share price.

Earnings that are not paid out as dividends are called retained earnings. Management has two uses to which such retained earnings can be put. One is the purchase of either real or financial assets. As long as these assets are productive, their acquisition will increase future earnings and hence the future price of the shares. The second use of retained earnings is the purchase of the firm's own shares in the open market. Reducing the number of stocks will also increase future per-share earnings and dividends and hence the future price of the stock. Therefore if the firm buys productive assets or buys its own stocks, per-stock earnings and dividends will increase.

Table 4.4 shows that the average US dividend yield from 1927 to 1995 was about 4.4%.

Table 4.4 US dividend yields, returns and dividend growth (% per year), 1927–95

	1927–95	*1927–45*	*1946–95*	*1927–71*	*1972–95*
Dividend yield	4.37	4.95	4.14	4.52	4.14
Real returns	6.43	4.86	7.03	6.73	6.40
Real dividend growth	1.12	–1.62	2.15	0.71	2.34

Source: Cochrane, 1997b

Apart from the historic low of the dividend yield, two other questions need to be asked when using this as a form of calculating investment returns.

First, how safe is the dividend yield? If the next dividend payment is cut, the apparently attractive yield of, say, the real 7% return for 1946–1975, may be largely an illusion. Usually the market's expectation of the next or *prospective* annual total dividend is more important for the stock price than the amount actually paid in the past year. Stability of profits, and the dividend *cover* – the ratio of available earnings to dividends – are also important factors here. Secondly, how rapidly is the dividend likely to grow? A fast growing dividend stream is clearly worth more than a static one, as we discussed earlier, with the effect that growth stocks are worth buying on much lower dividends than mature companies.

Professional analysts use the dividend discount models discussed earlier to compute the impact of different growth assumptions. Due to the power of compound arithmetic valuations can be very sensitive to the forecast of the growth rate and this explains why company shares can fall so sharply when they produce disappointing results or issue profit warnings.

Suppose that the end of the millennium low dividend yield adjusts back to its mean. In this case, either stock prices will fall or dividends will increase. On 14 March 1998, Warren Buffet, a famous value investor, told stockholders that he believed the historically high stock prices were justified as long as interest rates remain low and corporations continue to produce 'remarkable' returns on equity. Buffet apparently believed that future dividends will rise to justify the low dividend to price ratio. Unfortunately history tells a different story. It is adjustments in stock prices, not dividends, that historically have driven the dividend yield back toward its mean value. If dividend yields do revert to their long-run mean over the next few years, the long-term prospects for the stock market change dramatically.

Unfortunately the analysis of dividends has become more complex than it used to be. Tax considerations cloud the calculations. Companies are therefore much more interested than they used to be in finding alternative, more tax-efficient, ways of returning cash to shareholders. Stock buybacks are increasingly popular. Analysts now agree that it is necessary to take stock buybacks into account in valuing dividend flows.

Bond yield/stock yield ratio

Investors always have a choice between stocks and bonds. In the long run, the two asset classes ought to achieve similar returns, or at any rate similar returns after adjusting for differences in risk and liquidity. In the UK it is usual historically for government bonds to pay a little more than twice as much as dividend yields. In the US the historical average is nearer three.

Analysts use the bond/stock yield ratio as a benchmark as to whether bond or equity prices will revert back to the historical average. Table 4.5 illustrates how this would work.

Table 4.5 Why bond yields drive stock (dividend) yields (and vice versa)

	Expectation				
Historical average bond yield/stock yield	*Assumed current ratio*	*Share price*	*Dividend yield*	*Bond price*	*Bond yield*
3	<3	↑	↓	↓	↑
3	>3	↓	↑	↑	↓

However, the dividend yield has to be sensitively interpreted. If the dividend yield is low due to the effect on earnings of a forthcoming recession then the ratio has to be treated cautiously.

Tobin's q ratio

In addition to comparing a company's dividends or earnings to the price of the stock it can also be useful to compare the value of the company's securities to the value of its assets. The ratio of the market capitalization to the replacement cost of the underlying corporate assets is known as Tobin's q. It is named after its creator, James Tobin, Yale University's Nobel Laureate.

Tobin's q is a simple enough idea. The ratio of (a) the value of companies according to their market capitalization to (b) the replacement of the cost of their assets should be a number equal to 1.The numerator and the denominator of the ratio are, after all, just two ways of measuring the same thing, the value of companies.

The value of the company on the financial markets exceeds the value of its assets when its expected earnings are high or rising rapidly. Conversely the company can be worth less than its assets when its prospects are especially poor or uncertain.

If q was for some reason greater than 1 this would imply that the stockmarket put a higher value on assets owned by companies than those same assets cost to acquire. So there would be a kind of arbitrage opportunity. If q were greater than 1 companies could be expected to increase their investment, presumably until the market reckoned the company was now worth what its assets cost. At that point q would again be equal to 1.

Estimates of Tobin's q for non-financial corporations by the US Federal Reserve, for 1998–2000, show the highest valuations since 1925. Using the Fed's calculations it has fluctuated between a minimum of 0.4 and a maximum of 1.8, its level at the end of the millennium. The average level is 0.7.

A value higher than unity, the case since 1991, for the first time in 25 years, indicates that it is more advantageous for firms to issue shares in order to finance the accumulation of productive capital (land, plant) than to buy such assets 'second-hand' through purchasing the corresponding securities on the stock market.

In the early 1970s and early 1980s the value was usually less than 1, suggesting that companies were quite cheap. If this ratio was to revert to the long-term average the US stock market would decline by 50%.

Smithers and Wright (2000) have recently analysed the US stock market and come up with a q ratio of well over 2. They find that when the ratio (suitably adjusted to take account of accounting discrepancies) moves far above unity, equilibrium is indeed eventually restored – not however by a surge in the replacement value of companies' assets but by a dramatic fall in their stock price. As Smithers and Wright point out, that is what happened in 1968–74 after earlier peaks in q. The value of q in year 2000, they note, is higher now than in either 1929 or 1986.

A ratio of this type has to be handled with great care, because of the problems posed by the valuation of assets both tangible and intangible. Intangible assets (intellectual property, trademarks, patents, know-how,) for example tend to be undervalued in the denominator of q, causing the ratio to be overstated. This argument is often advanced to explain the present high level of q, on the grounds that intangible assets are now of much greater importance than they were. We return to these ideas in Chapter 11.

Price to book value (P/BV)

Company balance sheets display a figure for the book value of stockholders funds, which can be expressed as net worth per share. US stocks in year 2001 were selling at a record 4.9 times book value. Book value refers to the value of the assets on a company's balance sheet. In year 2001 it was about 2.5 times the average multiple over the past 20 years.

But there is an old investment saying that 'assets are only worth what they can earn'. Asset analysis is most useful when companies possess unexploited assets that could be sold or put to other uses – office blocks, say, or stakes in other companies. Inflexible assets such as machinery or oil wells may need to be analysed more cautiously.

The decline of manufacturing industry has highlighted, as mentioned above, the importance of intangible assets in today's companies – brand names, software, royalty rights and the skills of the workforce. Because it is so hard to value these satisfactorily in the balance sheet the use of asset value analysis, and with it price to book value, has declined in importance.

Price to sales ratio (P/SR)

This is simply the ratio of the value of the company's stock (that is, its market capitalization) to its sales. This might seem a confusing indicator, because profit margins can vary so much between different kinds of industry. But in an influential book, *What Works on Wall Street*, published in 1996, the American quantitative investor James O'Shaughnessy described the price to sales ratio as the 'king of value factors'.

O'Shaughnessy found that low capitalization-to-sales ratios stocks beat the market more than any other value ratio. He stressed that investors should be cautious of stocks where the market capitalization is more than twice the level of the company's sales.

Using data going back to 1951 O'Shaughnessy analysed all of the basic investment strategies used to select common stocks, such as book value, cash flow, P/E, ROE, yield, and so forth. O'Shaughnessy found that the 50 stocks with the lowest PSRs based on an annual rebalancing of the portfolio, performed at an annual rate of 15.42% over the 40 years since 1954 through 1994, compared to 12.45% annually for his universe of all stocks. Stocks with the highest PSRs earned only 4.15% annually. Furthermore, combining low PSR stocks (generally, a PSR of 1.0 or lower) with stocks showing *momentum* (the best 12-month price performance) produced results of 18.14% annually over the full 40-year period.

Economic value added (EVA)

A different approach to valuation turns the whole issue on its head and is on the frontier of investment analysis. Instead of forecasting all the inputs to come up with a number for what the stock should be worth, this method starts with the stock price and works backwards to answer the question: what kind of growth does this company have to deliver to justify this price? This method does not avoid making some assumptions, but it's a way to perform a reality check.

Look at Amazon.com. At a price of $214 before the stock price collapse, the market was implying that Amazon's revenues will increase 59.6% a year over the next 10 years. That was the conclusion of veteran securities analyst Charles R. Wolf of Warburg Dillon Reed.

To come to that conclusion, Wolf used a valuation method built on economic value added (EVA), a concept developed by Stern & Co., a management consulting firm. The idea behind EVA is that in the long run, it's not accounting profits – taking in one more pound than you put out – but economic profits that matter. Simply put, a company earns an economic profit only if it has earned more than its cost of capital, something which is not found on an income statement. EVA was originally designed more as a management tool rather than an investing tool.

The idea in EVA analysis is that the market value, the stock price times number of shares, has two components. One, the current operations value (COV), is a measure of the worth of the company as it now operates. The second, future growth value (FGV), measures the company's expected growth. Once you determine the COV (that's the easier of the two) you can figure out the implied future growth value. And once you know that, you can determine the implied revenue growth rate. Then you can make a judgement as to whether that growth rate is achievable.

Critical to the analysis is the cost of capital. For Amazon.com, Wolf estimated a cost-of-capital charge of 15% – a figure derived from such variables as the risk-free rate of return, the extra return that equities historically return over bonds, and Amazon's 'beta', or price volatility as compared with the stock market.

Can Amazon.com. achieve a nearly 60% average annual revenue growth rate over 10 years? That's where investors must turn to industry fundamentals and old-fashioned common sense. For instance, using Wolf's calculations, Amazon should reach sales of $63 billion in 10 years. 'Is that realistic?' is the question investors need to ask.

5

Applying the Porter model to the valuation of Internet and technology stocks

The recent performance of technology and Internet companies and their stock price behaviour offers a classic textbook example of how business strategy, as taught in most business schools, can be applied. As any new industry establishes itself, the theory goes, it will move through several stages. First, many industry players will emerge. Second, a group of leaders will start to stand out from the pack. Third, these leaders will then move to cement their positions through acquisitions and consolidation, resulting in a divergence between the stronger and the weaker players, with the weaker players then dropping out. Eventually, a new group of more specialized, niche-oriented companies emerge to coexist with the leaders.

The textbook model demonstrates that one can apply this framework for analysing New Economy companies and ultimately their stock values in ways similar to the factors influencing old economy companies and their stock values. Analysis of the factors internal and external to the firm has to take place. New economy, old economy, same process. It is these internal and external factors that will influence the long-term valuation of these companies and it is to this that we now turn. Remember that the rules of the New Economy from Chapter 2, despite recent volatilities, must still be imposed on the standard valuation process. But what are these internal/external factors which need to be analysed? Internal factors are those which arise from within the individual company and external factors are those relevant to the industry overall.

Internal analysis: strategic factors within the company

The internal factors that are crucial to make a company successful, and ultimately affect its stock valuation, are listed below.

Human capital

- *Management:* Most successful companies have a strong management or a strong leader. A company needs to have a management that is able to motivate employees so they can work to their best ability. Investors consider management as a very important factor when they consider investing in a company; few investors would invest in a company with weak management.
- *Motivation:* It is important to keep all people involved in the company's activities motivated to work well. It has been very common among Internet companies to use stock options where the employees have the opportunity to have ownership in the company and thereby have a share of future profits. Clearly this becomes a problem when the stock price falls and the options appear to have little value.
- *Knowledge/skills:* Another important factor is to see what access to knowledge the company has. A company with high level of knowledge is probably more attractive than a company with a brilliant business idea but without the skills to carry it out. Initially attracting individuals with the appropriate skills is essential, but as the sector matures one would expect the supply of suitably skilled personnel to rise.

The corporate culture

The culture within the company is of great importance. Is the firm profit oriented, cost oriented? Do the employees have the right attitude? Without attributes like those the firm will not have the requirements to be a successful company. The work environment could be of considerable importance. Is the environment a creative environment to work in? Is the organization flexible? Do the employees feel that they are important and what they say counts? There is no doubt that the corporate culture for technology companies is very different from the old economy companies.

The uniqueness of product or service

Another important factor to analyse is the uniqueness of the product or the service. A good way to measure uniqueness is to estimate the time it would take a competitor to reach the same position the company has now. The most interesting companies have products or services that are protected by patents and/or are technically advanced or are in some other way difficult to substitute. Looking at a company's Research & Development (R&D) department will give a good view of their ability to come up with new unique products. The brand is another factor that an investor could look at to see if the company has or is building a strong brand that will add value to the company. The whole B2B and B2C sectors are providing a unique, if yet unproven business model.

External analysis: strategic factors for analysing the industry environment

An industry analysis usually begins with a general examination of the forces influencing the organization. The objective of such a study is to use this to develop the *competitive advantage* of an organization in its ability to defeat its rival companies.

This type of analysis is often undertaken using the structure proposed by Porter. His basic model is illustrated in Figure 5.1. This is often called *Porter's Five Forces Model* because he identifies five basic forces that can act on the organization:

- the bargaining power of suppliers
- the bargaining power of buyers
- the threat of potential new entrants
- the threat of substitutes
- the extent of competitive rivalry.

The objective of such an analysis is to investigate how the organization needs to form its strategy in order to develop opportunities in its environment and protect itself against competition and other threats.

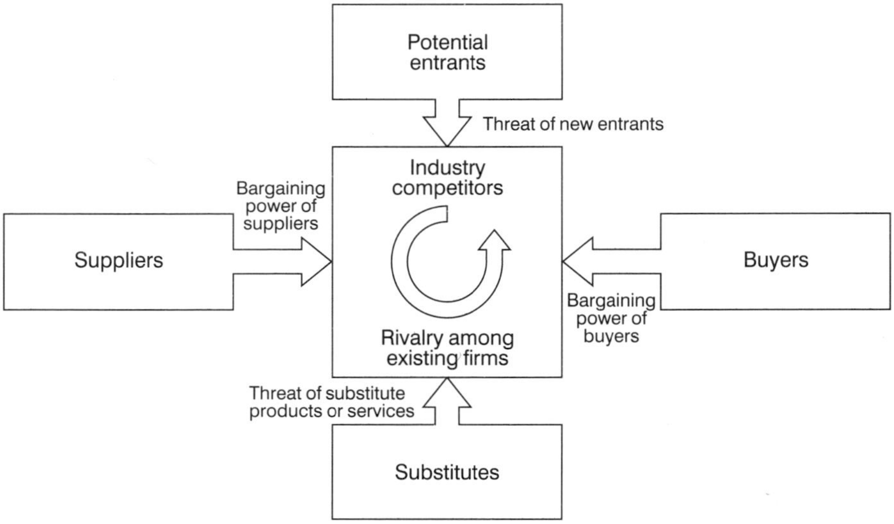

Figure 5.1 Porter's Five Forces Model. *Source:* Porter, 1985

The bargaining power of suppliers

Virtually every organization has suppliers of raw materials or services, which are used to produce the final goods or services. Porter suggested that suppliers are more powerful under the following conditions:

(1) *If there are only a few suppliers.* This means that it is difficult to switch from one to another if a supplier starts to exert its power.
(2) *If there are no substitutes for the supplies they offer.* This is especially the case if the supplies are important for technical reasons – perhaps they form a crucial ingredient in a production process or the service they offer is vital to smooth production.
(3) *If suppliers' prices form a large part of the total costs of the organization.* Any increase in price would hit value added unless the organization was able to raise its own prices in compensation.
(4) *If a supplier can potentially undertake the value-added process of the organization.* Occasionally a supplier will have power if it is able to integrate forward and undertake the value-added process undertaken by the organization; this could pose a real threat to the survival of the organization.

The bargaining power of buyers

In his model, Porter used the term *buyers* to describe what might also be called the *customers* of the organization. Buyers have more bargaining power under the following conditions:

- *If buyers are concentrated and there are few of them.* When the organization has little option but to negotiate with a buyer because there are a few alternative buyers around, the organization is clearly in a weak position: national government contracts in defence, health and education are obvious examples where the government can, in theory at least, drive a hard bargain with organizations.
- *If the product from the organization is undifferentiated.* If an organization's product is much the same as that from other organizations, the buyer can easily switch from one to another without problems. The buyer is even more likely to make such a shift if the quality of the buyer's product is unaffected by such a change.
- *If backward integration is possible.* As with suppliers above, the buyer's bargaining power is increased if the buyer is able to backward integrate and take over the role of the organization.
- *If the selling price from the organization is unimportant to the total costs of the buyer.*

The threat of potential new entrants

New entrants come into a market place when the profit margins are attractive and the barriers to entry are low. The allure of high profitability is clear and so the major strategic issue is that of barriers to entry into a market.

Porter argued that there were seven major sources of barriers to entry:

- *Economies of scale.* Unit costs of production may be reduced as the absolute volume per period is increased. Such cost reductions occur in many industries and present barriers because they mean that any new entrant has to come in on a large scale in order to achieve the low cost levels of those already present: such a scale is risky.
- *Production differentiation.* Branding, customer knowledge, special levels of service and many other aspects may create

barriers by forcing new entrants to spend extra funds or simply take longer to become established in the market.

- *Capital requirements.* Entry into some markets may involve major investment in technology, plant, distribution, service outlets and other areas. The ability to raise such finance and the risks associated with such outlays of capital will deter some companies. For many technology companies these cash requirements are very high, hence the use of the term 'cash burn' to measure how long they can stay in business before financial problems emerge.
- *Switching costs.* When buyers are satisfied with their existing product or service, it is naturally difficult to switch those buyers to a new entrant. The cost of making the switch would naturally fall to the new entrant and will represent a barrier to entry.
- *Access to distribution channels.* It is not enough to produce a quality product; it must be distributed to the customer through channels that may be controlled by companies already in the market. The widespread use of the Internet has made distribution a relatively simple exercise for technology-related companies.
- *Cost disadvantages independent of scale.* Where an established company knows the market well, has the confidence of major buyers, has invested heavily in infrastructure to service the market and has specialist expertise, it becomes a daunting task for new entrants to gain a foothold in the market. Whether or not these are effective barriers to entry in the technology sector is an unresolved question, particularly given the legal actions that Microsoft has been faced with.

The threat of substitutes

Occasionally, substitutes render a product in an industry redundant. But more often substitutes do not entirely replace existing products but introduce new technology or reduce the costs of producing the same product. Effectively, substitutes may limit the profits in an industry by keeping prices down.

From a strategy viewpoint, the key issues to be analysed are:

- the possible threat of obsolescence
- the ability of customers to switch to the substitute
- the costs of providing some extra aspect of the service that will prevent switching

- the likely reduction in profit margin if prices come down or are held.

The extent of competitive rivalry

Some markets are more competitive than others. In highly competitive markets, companies engage in regular and extensive monitoring of key competitor companies. For example:

- examining price changes and matching any significant move immediately
- examining any rival product change in great detail and regularly attempting new initiatives themselves
- watching investment in new competing software systems and having regular drives to reduce their own costs' levels
- attempting to poach key employees.

The factors that will lead to high competitive rivalry are:

- *When competitors are roughly of equal size and one competitor decides to gain share over the others,* then rivalry increases significantly and profits fall. In a market with a dominant company, there may be less rivalry because the larger company is often able to stop quickly any move by its smaller competitors.
- *If a market is growing slowly and a company wishes to gain dominance,* then by definition it must take its sales from its competitors – increasing rivalry.
- *Where fixed costs of storing finished products in an industry are high,* then companies may attempt to gain market share in order to achieve break-even or higher levels of profitability.
- *If extra production capacity in an industry comes in large increments,* then companies may be tempted to fill that capacity by reducing prices, at least temporarily.
- *If it is difficult to differentiate products or services,* then competition is essentially price-based and it is difficult to ensure customer loyalty.
- *When it is difficult or expensive to exit from an industry* (perhaps due to legislation or redundancy costs or the cost of closing dirty plant), there is likely to be excess productive capacity in the industry and thereby increased rivalry.
- *If entrants have expressed a determination to achieve a strategic stake in that market,* the costs of such an entry would be relatively unimportant when related to the total

costs of the company concerned and the long-term advantages of a presence in the market.

The Porter model, the Internet and the Web

The goal of any business strategy is to achieve a competitive advantage within the company's industry. The Porter model demonstrates that companies competing in a particular, established industry are interested in positioning themselves in the market but, above all, share a common interest in not letting anybody else in (to prevent the competition from growing stronger); in preventing new products or services that could replace their own from appearing; and, lastly, in ensuring that their internal competitiveness does not give rise to an increase in suppliers' or clients' negotiating power. It is by controlling all these factors that a company may obtain a competitive advantage within its industry.

The Internet is merely a physical medium that provides a base for certain services, but, as will be seen below, does have major implications for applying the Porter model. The main services are e-mail, newsgroups and the Web. The possibilities offered by each are very different, as are the rules governing them.

- e-Mail allows fast, inexpensive, and asynchronous communications. (It is not necessary for the two people communicating to do so at the same time.) In this regard, it has merits that the telephone (synchronous), the fax (expensive) and the postal service (slow) all lack. It is personal and, therefore, discrete (provided that nobody with sufficient technical skills tries to intercept it).
- Newsgroups are a means for receiving information on specific subjects that interest the group and are a means of disseminating this information. Their merit lies in the fact that the information is unrestricted and is not screened by the major communications corporations.
- Finally, there is the World Wide Web. The Web has broken through the barriers of the not-so-friendly Internet and has made it a user-friendly tool. It has transformed an information network into a real information channel that is open to the public in general. The public can now be seen to be potential buyers of whatever products are on offer.

In short, the Internet has created a new distribution channel. This necessarily provides sellers with a means of informing potential customers about their supply of products and enables customers to make their purchases (order and payment). It is the combination of the Web and e-mail which can provide companies with a full distribution channel. The Web is used to inform clients and e-mail to place orders. To simplify the process, e-mail can be generated directly from the Web.

As a distribution channel, the Web may be characterized as follows:

- It is a new channel that is not controlled by the companies from any one particular industry.
- Distribution costs are low.
- Its impact is limited (for the time being). Predictions are made, but its true impact is not really known.
- It enables fast distribution of information products.
- The cost of changing prices or even commercialized products is very low, thus enabling a relatively quick product rotation.

The power of the Internet is that due to the real time nature of the Internet, customers are able to get instant feedback about information or availability and delivery status of products. Its implications for the Porter model are discussed below. One of the great strengths of the use of the Internet is that it is independent of time, language and the geographic location of customer/company. It helps to break down traditional barriers to entry, shifts the power balance between buyers and sellers, facilitates comparison-shopping, distributes knowledge, expands markets and sustains growth.

The key features of the Internet for business are listed in Table 5.1.

The Porter model revisited

The arrival of the Internet has major implications when applying the Porter model (see Figure 5.2). The five competitive strategies have to be carefully analysed.

- The power shift between suppliers and buyers: the power of suppliers is no longer the same. The buyers can more easily

Table 5.1 Key features of the Internet for business

Key feature	*Implication*
Availability	'Normal' business hours time constraints do not exist any more with on-line, 24-hour per day and immediate access
Ubiquity	For most companies and customers there is no reason not to have Internet access – so assume that in the near future most will have it, just as we now assume that they have a phone and a fax
Global	With no physical borders for access (but not always for delivery), our mental map of what is 'near' and 'far' will radically change
Local	Paradoxically, the Internet not only enables global commerce, but is also a perfect vehicle to reinforce local physical presence and local person-to-person business relationships
Digitization	The real business will increasingly be happening in information space, even for physical products. Convergence is leading to the shake-up of major industries such as telecommunications and broadcasting, as well as different thinking about the 'natural' laws of economics, e.g. increasing rather than decreasing returns to scale
Multimedia	A long-awaited combination of technologies is finally coming to business, not only to gain a competitive edge in information provision during buying and selling but also to provide completely new opportunities in consultancy, design and entertainment in combination with interactivity and networking
Interactivity	Interactivity is a challenge, namely to overcome the virtuality of the business relationship, as well as an opportunity for greatly improving traditional customer service at an affordable price
One-to-one	Based on data processing and customer profiling, possibly based on enriched interactivity, one-to-one marketing is a natural companion of doing business on the Internet. It may also be a necessity to overcome the anonymity of Internet business relationships
Network effects and network externalities	Low cost and fast growth in the number of relationships enable business models that require a significant number of parties in the network and whose benefits increase faster with a growing number of parties. That is, these business models exhibit network effects and/or network externalities. Examples are implementations of open market concepts such as virtual communities and third-party market places
Integration	It has long been argued and to some extent demonstrated that the value of combined information across steps of the value chain is more than the sum of its parts. The Internet now provides at least part of the technology for value chain functional and information integration. Advanced electronic commerce companies show how to exploit the added value

Source: Timmets, 1999

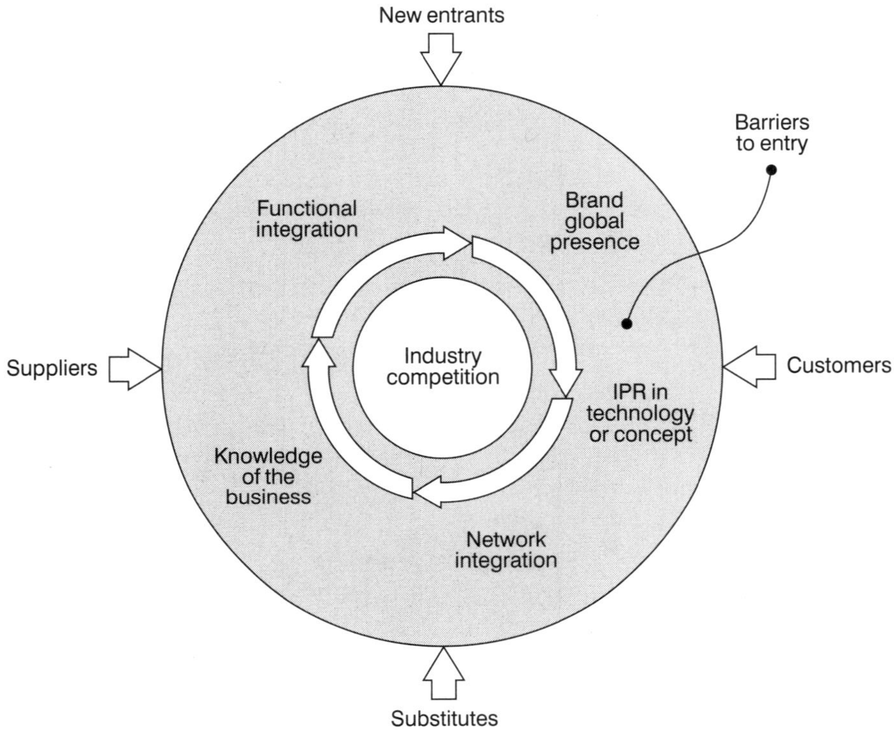

Figure 5.2 Competitive forces and barriers to entry (the Porter model) applied to the Internet

access other suppliers that may offer better conditions for the same product.

- The barriers to entry: the usual high barriers to entry for certain businesses are coming down. Amazon demonstrates perfectly that barriers to entry are coming down with its virtual bookstore, providing more choice than the great majority of stores.
- Competitors and substitute suppliers: the lack of boundaries in the Internet guarantees that substitute suppliers and the competitors are abundantly available.

Internet electronic commerce brings the textbook ideal of perfect competition closer in the sense that:

- barriers to entry are lowered
- transaction costs are reduced
- customers have improved access to information
- marginal or customer-orientated pricing becomes possible
- minimal legislation and regulation and other forms of intervention by public authorities are now the norm.

Industry competition can be expected to grow as service providers are increasingly implementing the same generic concept. Specific characteristics of the Internet in relation to industry competition revolve around an increase in suppliers' power.

Suppliers' power

The suppliers of the Internet are technology and platform providers (Yahoo, Microsoft, etc.). The objective of platform providers is to establish and control the dominant standard. This strategy is based, for example, on the success of Microsoft MS DOS as the de facto standard operating system of the 1980s and Windows for the 1990s, control of which gave Microsoft enormous leverage in application software.

The market dominance strategy of the technology providers, such as Microsoft, means that technology users can obtain major parts of the basic technology rather inexpensively. On the other hand, those technology users cannot derive strategic advantages from the basic technology, as it is generally available. They may be in a strong position to negotiate especially advantageous price deals, as the technology providers are strongly interested in enhancing market share. In the future the situation might change, if relatively small companies with particular circumstances supply critical technology for highly specialized advanced services, for example technology for intelligent service assistance or one-to-one marketing.

The reaction to this threat to existing businesses has been defensive. In particular large companies have reacted to it by applying familiar concepts of branding, to which we now turn.

Technology stocks and branding

Branding is a major factor in product strategy decisions for technology companies. But what is a brand and why is it so important for technology companies? The origins of the word 'brand' go back to old English, where it meant a burning piece of wood. Over the years it came to mean something that was created as result of intense heat, associated with branding criminals or cows.

The American Marketing Association defines a brand as

> A **brand** is a name, term, sign, symbol, or design, or a combination of them, intended to identify the goods or services of one seller or group of sellers and to differentiate them from those of competitors.

Interbrand, a specialist brand consultancy, have a more financially specific definition of a brand. Interbrand regards the brand as an intangible asset that creates an identifiable economic earnings stream. Their approach to brand valuation is based on the principle of economic use, which means the value the brand adds to the underlying business. Brand value is defined as the net present value of the economic profit that the brand is expected to generate in the future. Table 5.2 illustrates that many of the world's most well-known companies have a brand value, rather than an asset value, which represents a significant proportion of their market capitalization.

In essence, a brand identifies the seller or maker. It can be a name, trademark, logo or other symbol. Under trademark law, the seller is granted exclusive rights to the use of the brand name in perpetuity. Brands differ from other assets such as patents and copyrights, which have expiration dates.

A brand is essentially a seller's promise to deliver a specific set of features, benefits and services consistently to the buyers. The best brands convey a warranty of quality. But a brand is an even more complex symbol. It can convey up to six levels of meaning:

(1) *Attributes*: A brand brings to mind certain attributes. Mercedes suggests expensive, well-built, well-engineered, durable, high-prestige automobiles.
(2) *Benefits*: Attributes must be translated into functional and emotional benefits. The attribute 'durable' could translate into the functional benefit: 'I won't have to buy another car for several years.' The attribute 'expensive' translates into the emotional benefit: 'The car makes me feel important and admired.'
(3) *Values*: The brand also says something about the producer's values. Mercedes stands for high performance, safety, and prestige.
(4) *Culture*: The brand may represent a certain culture. The Mercedes represents German culture: organized, efficient, high quality.

(5) *Personality*: The brand can project a certain personality. Mercedes may suggest a no-nonsense boss (person), a reigning lion (animal), or an austere palace (object).
(6) *User*: The brand suggests the kind of consumer who buys or uses the product. One would expect to see a 55-year-old top executive behind the wheel of a Mercedes, not a 20-year-old secretary.

In the digital world consumers are in control. Digital developments allow Web surfers to choose products solely on real value and not just intangibles, like brand associations. Already websites exist that compare features on more than 10 000 products and digital personal assistants will surf the Web looking for information you've requested. This development is in fact weakening the impact of brands.

Despite this long-term trend towards brand weakening, how does a small company or a new company which cannot spend millions of dollars on an expensive advertising campaign compete? Technology companies in particular have been adept at achieving levels of brand recognition through less conventional marketing approaches. Here are two examples:

American Online Inc.

Over half of US households are familiar with America Online (AOL). That's because AOL gives away its software for free. For several years AOL has been blanketing the country with diskettes and now CD-ROMs, offering consumers a one-month free trial. Because it's hard to describe the benefits of an on-line service to novices, AOL believes the best approach is to let them try it. Then, once consumers start using AOL, the company reasons that the user-friendly program will lure them to subscribe. Also on AOL's side is sheer consumer inertia, which prevents many people from switching and signing on with another Internet service provider.

Sun Microsystems Inc.

Sun has built the visibility of Java, its flagship software program, within the corporate community almost entirely through conducting a public relations war. It has aimed its guerrilla PR largely at the company's nemesis, Microsoft. For example, after Microsoft developed a technology that competed with Sun's 'Java Beans', Sun learned when and where Microsoft planned to introduce it. The day before Microsoft's announcement Sun mailed bags of coffee beans to reporters with a note saying 'Why is Microsoft so jittery?' and inviting them to attend a Sun training seminar on Java Beans at a hotel next to where Microsoft was holding a developers' conference and making its announcement. Sun says the tactic was a hit: It attracted over 250 people and planted seeds of doubt in reporters' minds about the Microsoft technology. 'Think of it as a military operation', said John Loiacono, Sun's vice president of brand marketing.

Several of the most valuable brands have a tiny (or in some cases virtually zero) tangible asset base: Yahoo, Amazon and many of the Internet based companies (see Table 5.2). In many of these cases the company's *only* asset is its brand. So a brand is a 'relationship', a 'reputation', a set of 'expectations', a promise. Even allowing for stock market fluctuations

Table 5.2 The world's most valuable brands by brand value

	Brand	*Brand value ($US m)*	*Market capitalization of parent company ($US m)*
1	Coca-Cola	83 845	142 164
2	Microsoft	56 654	271 854
3	IBM	43 781	158 384
4	General Electric	33 502	327 996
5	Ford	33 197	57 387
6	Disney	32 275	52 552
7	Intel	30 021	144 060
8	McDonald's	26 231	40 862
9	AT&T	24 181	102 480
10	Marlboro	21 048	112 437
11	Nokia	20 694	46 926
12	Mercedes	17 781	48 326
13	Nescafé	17 595	77 490
14	Hewlett-Packard	17 132	54 902
15	Gillette	15 894	42 951
16	Kodak	14 830	24 754
17	Ericsson	14 766	45 739
18	Sony	14 231	28 933
19	Amex	12 550	35 459
20	Toyota	12 310	85 911
21	Heinz	11 806	18 555
22	BMW	11 281	14 662
23	Xerox	11 225	27 816
24	Honda	11 101	30 050
25	Citibank	9 147	42 030

Source: Interbrand/Citibank, 1999

the value of many of the technology companies is still far higher than conventional accounting practices would deem appropriate.

Financial analysis in the valuation of companies

In analysing the financial position of a company, technology or otherwise, it is essential to assess both qualitative and quantitative aspects of the company (see Table 5.3).

Qualitative company analysis

Qualitative analysis would include:

- an individual overview of the business
- an overview of the industry
- the position of the business within the industry
- a management profile
- analysis of 'event risk' – the risk that a single event could fatally affect the company.

Quantitative company analysis

Quantitative analysis would include:

- profit and loss analysis
- balance sheet analysis
- operational leverage
- break-even analysis.

All companies, technology companies included, seeking to raise finance on the international capital market are frequently obliged to gain a credit rating from one of the credit rating agencies. These agencies analyse the ability to repay any borrowings by analysis of the balance sheet, profit and loss account and the cash flow statement. The criteria applied by the rating agency Standard & Poors provide a valuable guide as to which financial information should be analysed when analysing the financial strength of any company (see Table 5.4).

Table 5.3 Qualitative company analysis

Business overview	
Company history	Where has the company come from and where is it going? This is not only about historic financial performance, but also longevity and culture (e.g. has the company grown organically or via acquisition?). If a company has survived historic economic fluctuations, changes in technology, culture and the outside world, it is fair indication that the company will survive in the future, although this is not guaranteed
Structure	What is the company structure? How are subsidiaries controlled? Where are the assets located/situated (i.e. can they be repatriated) and where is the cash generated?
Shareholders	Who are the main shareholders, what is their interest/purpose in owning the company? Could they offer support?
Industry analysis	
Political/ economic importance	Is there political risk? Does the industry have national, political or economic importance? Is it likely to receive government support if necessary?
Regulatory risk	Is the regulatory environment supportive or hostile? Could this change? What would the implications be?
Size/growth potential	What are the industry prospects? Is it growing or declining? Are there any technological changes on the horizon, a very important consideration with technology stocks? Is it possible to raise prices? Are there any substitute products? Is the industry consolidating?
Competition	What is the competitive structure of the industry? Is the industry a monopoly, duopoly, oligopoly or is it highly competitive? Is there pricing flexibility?
Barriers to entry	Are barriers to entry a competitive advantage, or will they work against the company?
Profitability	Is the industry profitable? A company can be very efficient and well run, but if the structure of the industry is such that it is unprofitable, it may not make money
Cyclicality	How resilient is the industry to an economic downturn? How volatile has past performance been, again very important for technology companies? How does cyclicality feed through to profit margin and cash flow?
Key success factors	Is the industry driven by price, quality or service, quality of product, distribution capabilities or other factors? Is its success placed on technology developments only? Why should the customer buy this company's product?
Business position	
Country/political risk	The economic, political and regulatory conditions of the company's home country and/or places of doing business are an important starting point, i.e. what is known as a 'top-down' approach

Table 5.3 Continued

Size	How big is the company compared to its competitors, suppliers and buyers? Size is important because it usually implies either diversification or power. Also, large firms can normally survive for a period even if their performance suffers, particularly true for technology companies with low earnings
Product mix/ diversification	Are the products new or tested? Are the products/services value-added or commodity? What is the percentage of R&D expenditure? What is the spread/mix of products? Single product companies are normally more risky, although this has to be balanced against diversification within the product range and the spread of buyers and suppliers
Demand/supply characteristics	Does the company have contracted revenues that effectively guarantee a minimum level of cash flow? Is there a forward order book? Is it growing? What is the companies bargaining power in relation to suppliers and buyers? Are there alternatives?
Labour relations	Are strikes/disruptions likely? What would be the impact? How easy is it to reduce staff numbers and cut costs?
Customer profile	Who are the customers? Why do they buy the company's product/services? How elastic is their demand? How price-sensitive are they? Does the company continue to win new customers? Is it over-reliant on any one customer in particular?
Geographic diversification	What is the geographic split? Will this cushion or exacerbate a slow down in demand? Geographic diversification can lend support by diversifying income into different markets so that cyclicality in one area is offset by growth in another
Production/ technology requirements	Is the company vulnerable to technological changes? Does it have to invest in R&D? Is the business capital-intensive?
Marketing and distribution channels	How effective and necessary is the marketing and distribution? How much does the company have to spend on this? This is a very important factor for technology companies
Management profile	
Philosophy and experience	Who are the management? How long have they been with the company and what is their track record? How has the company performed compared to its peers?
Industry knowledge	Are the management leading or following other industry players? Does this increase or decrease the business risk?

Table 5.3 Continued

Business plan	Is it realistic? Have they met plans in the past? Many failed Internet companies had poor business plans
Operating strategy	Are the management focused on market share and sales, or on margins? Are they focused on the short-term profit or longer-term stability? Do they have the vision to make the right strategic choices?
Financial strategy	This encompasses both accounting policies and management's attitude towards risk and debt. Are acquisitions planned or is the company focused on debt reduction?
Succession	Does one man dominate? Is there a successor? Is there a balanced team?
Event risk	
Management changes	Management changes, either following a boardroom coup in response to poor performance and shareholder pressure or a natural changeover following retirement, often signal the start of a new era and a change in strategy or direction. Management changes should be carefully monitored
Low growth	A mature cash generative business may appear a strong/stable company, but if the industry is not growing, management may be forced to change direction or diversify to boost growth prospects and satisfy shareholders. This may result in greater business and/or financial risks being taken
Share price under-performance	The management of a company are employed by the shareholders who own the company. Their chief responsibility is to run the business for the benefit of the shareholders by enhancing returns through higher dividends and an increase in the share price. A prolonged period of poor share price performance will almost inevitably have consequences, as the shareholders make known their dissatisfaction. This will normally result in either a change in strategy, a change of management, a more aggressive financial structure to increase returns, an acquisition to gain exposure to a higher growth area, or ultimately a loss of independence
Regulatory change	Regulatory changes can act as a spur to changes in an industry
Industry consolidation	Often one particular merger or acquisition in an industry will prove defining and eventually result in a change in the whole industry landscape. Once an industry starts to consolidate it becomes self-perpetuating. Each remaining company feels compelled to seek a partner in order to ensure critical mass and maintain its competitive position

Table 5.4 The eight key Standard & Poors financial ratios

S&P ratio	*Definition*	*Usage*
Pre-tax interest cover[1]	Pre-tax income plus interest expense less interest expense	This shows how many times profit covers the amount of interest due. The higher the figure the more the profit could decline before a company fails to pay its interest
EBITDA interest cover	(Pre-tax income plus depreciation plus interest expense less interest capitalized)/(Gross interest expense)	This shows how many times cash generated covers the amount of interest due and demonstrates the ability to generate cash to service debts. As above, the higher the coverage, the higher the margin of safety before a company fails to pay its interest
Funds from operations (FFO) as a % of total debt[2]	(Net income before extraordinary items plus depreciation plus deferred tax plus non cash items)/(Total debt) × 100	This shows how much cash the business generates after payment of interest and tax but before re-investment. The higher the percentage the less time it would take the company to repay all its debt. If a company has 100% FFO, then it could (in theory) repay all its debt within 1 year
Free cash flow as a percentage of total debt[3]	(Net income before extraordinary items plus depreciation plus deferred tax plus non-cash items less capex plus the increase in the change in working capital)/(Total debt) × 100	This shows how much free cash the business generates i.e. how much is available for debt repayment after re-investment in the business (i.e. working capital and capex). As above, the higher the percentage the shorter the time the company would need to repay its debt whilst maintaining current investment levels
Return on permanent capital	(Pre-tax income plus interest expense less interest capitalized)/(The 2 year average of total debt plus deferred taxes plus minority interests plus shareholders equity) × 100	This shows the return the business makes on the resources employed. If the return is not high enough there may be shareholders' pressure to improve it, increasing the risk of an acquisition or a share buyback, and increasing the risk of a bid
Operating income as a % of sales	(Operating income before depreciation)/(Sales) × 100	This shows how much profit the company makes from each sale. The higher this ratio, the more operating profit the company will generate as sales grow
Total debt as a percentage of total capital	(Total debt)/(Shareholders' equity plus total debt) × 100	This shows how much of the resources employed in the business are provided by debt. The higher the percentage the more debt in the business and the lower the 'cushion' of equity. This is important because whilst interest must be paid, in need a company could decline to pay a dividend
LTD as a percentage of total capital	(Long-term debt)/(Shareholders' equity plus long-term debt) × 100	As above, this shows how much of the resources are made up by LTD and therefore how much reliance the business places on short-term debt

[1] Pre-tax income as profit before tax, extraordinary items and minority interests.
[2] Net income is after tax and before dividends.
[3] Working capital excludes cash, cash equivalents and short-term debt. Capex is gross.
Source: Barclays Capital/Standard & Poors

6

Applying traditional valuation models to Internet and technology stocks: what are the problems?

The most common discounted cash flow (DCF) approach used in stock valuation is a two-stage model based upon enterprise (free) cash flows, as discussed in Chapter 4. The first stage is an explicit forecast, often for five years, of enterprise cash flows, and this is followed by a terminal value based upon assumed constant growth of the Year Five cash flow in perpetuity. While mathematically sound, there are two main practical problems with this approach, particularly as applied to Internet and technology stocks.

(1) The limited explicit forecast period fails to capture all the high growth. Many growth stocks have been traditionally priced on the assumption that a premium growth rate can be maintained for longer than, say a five-year explicit forecast. This means that the growth rate in the terminal value calculation has to be a quasi-average of continuing high growth for a limited period plus the real long-term growth; in other words, a number almost impossible to estimate.

(2) The final-year forecast cash flow is a poor basis for measuring terminal value. In traditional two-stage DCF models, typically 70–80% of value is inherent in the terminal value calculation, whereas for growth stocks this is usually well over 90%. For technology stocks it is not uncommon for over 100% of value to be in the terminal value (the cash flows in the explicit forecast period are negative).

Since the terminal value is based upon the cash flow in the final year of the explicit forecast period, this becomes the critical figure in the valuation. However, because a technology company has probably not reached maturity at this time, the basis of this cash flow forecast – the margin, capital expenditure, etc. – are not always appropriate.

One other major problem with DCF is that since it is largely based upon expected values it fails to recognize the value inherent in management's ability to change strategy when things do not go as well as expected and to take advantage of further opportunities as they arise. This brings us back to more subjective issues in valuing technology and Internet stocks such as the business model, flexibility and quality of management, as discussed in Chapter 5. In addition to these problems with DCF, the whole valuation of assets issue again reasserts itself.

Financial accounting and intangible assets

The accurate measurement of profit is fundamental to financial accounting. Profit tells us two things: how much revenue exceeded costs (a measure of the economic value of the current operations of the firm) and how much the assets of the corporation have increased (before any cash distributions to stock holders). Formally, accountants define profit as 'the excess of revenues over all expenses'. Expenses are 'the costs of goods, services, and facilities used in the production of current revenue' (Estes, 1981).

To the extent that a firm buys inputs that are not used up in production, those additional costs are investments, not expenses, and are capitalized, that is, considered assets. A capital asset gives rise to an expense only to the extent that the capital asset's value falls while in use, a process normally called depreciation or capital consumption. The intertwining of the measurement of corporate earnings and corporate assets depends on how one defines investment and assets. To understand how definitions of investment affect the measure of profit, which is very important in valuing technology stocks, we need to follow the details of corporate profit accounting.

The world according to GAAP

In the United States, corporate books are kept by certified public accountants who apply a set of rules called 'generally accepted accounting principles' (GAAP). According to GAAP, 'All R&D costs covered by GAAP are expensed when incurred', that is, R&D costs are treated as part of the current expenses of the firm and this treatment reduces reported profit. The only part of R&D costs not expensed is purchases of durable, tangible assets 'that have alternative future uses' beyond the project in hand. The rationale for this treatment of R&D is, in part, that firms might be tempted to artificially manipulate profits if R&D were capitalized. For example, by pretending that some ordinary expenses of the business were R&D, the firm might disguise a loss. Another part of the rationale is that R&D expenditures are more speculative than investments in fixed assets (fixed assets may have alternative uses and thus could be sold to others, but the product which the R&D is designed to exploit may not be successful and therefore have no alternative use).

Expensing versus capitalization of R&D expenditures

The reported earnings of technology firms are misleading, given the way in which they treat R&D expenses. Under the rationale that the products being researched are too uncertain and difficult to quantify, accounting standards have generally required that all R&D expenses are to be expensed in the period in which they occur. This requirement has several consequences, but one of the most profound is that the value of the assets created by research does not show up on the balance sheet as part of the total assets of the firm.

Many commentators have argued that research expenses not withstanding, the uncertainty about future benefits should be capitalized. To illustrate how the accounting principles influence the price/earnings ratio let us take a simple example. Assume in both examples below that revenues = $100, operating costs = $50, R&D expenditures are $20 and the company share price is $1000. As can be seen from Figure 6.1, if R&D is expensed the P/E ratio is 33.3 and, as can be seen from Figure 6.2, if it is capitalized it falls to 21.7.

In other words, by expensing rather than capitalizing R&D, P/E ratios are overstated. P/E ratios thereby have severe limitations when valuing technology companies. This has major

Profit and Loss A/C (end of t)

Revenue = \$100

Operating costs = \$50

R&D expenses = \$20

Earnings = \$30

P/E Ratio = $\frac{\$1000}{\$30}$ = 33.3

Balance Sheet (end of t)

Assets	Liabilities
Tangible Fixed Assets \$100	Equity \$100
Intangible Assets \$0	
Cash \$30	Reserves \$30
Net assets \$130	

Figure 6.1 R&D is expensed

Profit and Loss A/C (end of t)

Revenue = \$100

Operating costs = \$50

R&D expenses = \$4

Earnings = \$40

P/E Ratio = $\frac{\$1000}{\$46}$ = 21.7

Balance Sheet (end of t)

Assets	Liabilities
Tangible Fixed Assets \$100	Equity \$100
Intangible Assets \$16	
Cash \$30	Reserves \$46
Net assets \$146	

Figure 6.2 R&D is capitalized and amortized at 20%

implications when applying P/E ratios when making investment decisions, as will be seen below.

In the so-called New Economy, R&D, technical knowledge, brands etc. makes up an increasing share of a company's value. But they typically are missing from audited accounts that supposedly show how much a company is worth. It is this paradox which explains the high valuations that historically have been placed on these companies. Markets were valuing assets, so the story goes, that conventional accounting principles did not capture. As information becomes more valuable in the New Economy, and information is at the core of the value of intangible assets, perhaps, so it is argued, it is time to change the accounting arrangements. The problem still remains, however, of how to value knowledge-based assets. Is advertising an investment in building a brand an asset or is it just another cost of doing business? Should software development be counted as an investment, or should it be counted as an expense?

The role of intangible assets: if you cannot touch it what is it worth?

The word intangible is defined by *Webster's Dictionary* as something 'that represents value but has either no intrinsic value or no material being'. The fact that intangible assets are lacking in intrinsic value or material being makes it difficult to pinpoint their value, yet such assets often play an important role in the value of a business, a fact particularly true for technology companies. So how do you determine the value of intangible assets?

Every year, thousands of companies seek money from stock offerings, mergers, joint ventures and divestments. These are common financial techniques for either raising operating capital or for enabling the founders of a project to exit it. In all of these cases, the valuation of the company is a critical point. The valuation is the total worth of company's assets as agreed upon between a seller and a buyer(s).

Two types of assets influence the valuation of the company: tangible and intangible.

- **Tangible assets** are real-world, physical items. They include cash, real estate, machinery, equipment, inventory, etc.
- **Intangible assets** have value because of their implicit usefulness or worth. They include such things as proprietary knowledge, processes, and products, a desirable market segment, customer lists, and many other items.

For good reason, accounting practices tend to emphasize objective measures of value. In doing so, they treat the business as essentially an assembly of tangible tradable assets. Yet the real value in most businesses is in their intangible assets.

How much of the future revenue and profitability of a firm is due to its market share and the reputation it and its products now enjoy? How much is due to the cooperative and knowledgeable relations established over time with suppliers and distributors; to the expertise and insight of its personnel, and more particularly the ways they have developed integrated business approaches to produce products and serve markets; and to the systems and practices created within the firm as the way of doing business? The balance sheet statement is silent about these and other components of business upon which businesses depend for future earnings.

Types of intangible assets

Many businesses have intangible assets, and no list could ever be complete given that intangibles can be unique to a specific business. Some, however, are fairly common and can be found in many businesses. These more common intangibles include:

- **Proprietary lists:** Proprietary lists can include customer or client lists, patent lists or even mailing lists, whether they are made up of customers or prospects. Lists can be especially valuable to a business if the relationships they represent are ongoing. Consider, for example, a magazine's list of advertisers. The magazine may get 75% of its advertising revenue from the companies on that list. Therefore, the list is critical to the magazine's future profitability.
- **Beneficial contracts:** Long-term contracts can add value to a company. For instance, a company may have a contract that allows it to sell its product or service for a higher-than-normal mark-up. Or it may have a contract that allows it to purchase or lease items at a below-market rate.
- **Patents and applications for patents:** How much the patents are worth depends on the strength of the patent claim (which can be difficult to determine if the patent hasn't withstood litigation), and the patent's economic and legal life.
- **Copyrights:** Copyrights are trickier to value than patents because, while they may have a long legal life, their practical value may only be for a short period. This is especially true for technical works that become dated quickly. The value of a copyright also depends on the author's previous success.
- **Trademarks and brand names:** If a brand name or trademark enables a company to sell its products for a higher price or in greater quantity than its competition, it has value.
- **Subscriptions and service contracts:** Subscriptions are especially important for newspapers, magazines and cable companies because a large portion of their revenues is based on subscriptions. With both subscriptions and service contracts, the longer they have been in effect, the more they are worth.

- **Franchise agreements:** Franchises with long track records and well-recognized names have significant value over newer, lesser known franchises. This is especially true in some industries (such as the hotel industry) that are dominated by franchises.
- **Software:** Many companies have developed proprietary software specific to their businesses. If this software provides efficiencies and benefits that the business wouldn't have without the software, it is a separate asset.
- **Goodwill:** Goodwill means many things to many people, but generally it refers to intangibles like reputation and location that lead to repeat business.

Methods for valuing intangible assets

Valuing intangible assets can be done using one of three basic methods: cost of creation, capitalization of income or savings, or discounted cash flow. Rule-of-thumb formulas do exist for some assets in some industries but are better for providing an estimate of value to be compared with the value determined by the other methods.

Cost of creation

The cost of creation method of valuing intangible assets relies on calculating what it would cost another business to duplicate a given asset today. This method does not measure an asset's future impact on profits; it merely looks at what it would cost to create the asset from scratch at a particular point in time.

Assets that may be valued using the cost of creation method include:

- Internal software
- Patents
- Trademarks
- Copyrights
- Subscriptions
- Customer lists
- Service contracts

Capitalization of income or savings method

The capitalization method measures the future benefits intangible assets will bring to a company, identifies when those benefits will be generated and determines for how long. The capitalization rates used in this method should reflect the risk associated with the intangible asset being valued.

In addition to the income an intangible asset may bring to a company, the benefits may also include savings to the company as a result of owning the asset, or not having to pay a royalty to someone else who owns the asset or of efficiencies generated by the asset.

Intangible assets that this technique is suited to include:

- Trade names
- Customer lists
- Commercial software
- Patents
- Trademarks
- Brand names

The capitalization method works well for all of these assets when they are relatively new. As they come closer to the end of their economic usefulness, however, other methods of valuing them may become more appropriate.

Discounted cash flow

The discounted cash flow method is good for assets with predictable life spans and future financial benefits, including:

- Contracts (current and future yearly benefits)
- Subscriptions and service contracts
- Patent royalties

Methodology issues in technology stock valuations

Accounting issues are always important in stock valuation, particularly where valuations are based upon multiples of income statement data. We discuss the role of applying the concept of multiples in more detail in Chapters 8 and 9. The common accounting problems such as goodwill, stock options, in-process research and development, and pooling versus purchase accounting are all relevant to Internet and technology stocks as well as to other companies. However, these issues are reasonably well understood.

One of the features of multiple comparisons today is the use of measures from higher up in the income statement, such as cash earnings or EBITDA. The purpose of this is to mitigate the comparability problems caused by accounting differences through the elimination of goodwill amortization and depreciation, the accounting treatment of which can vary considerably, most especially where cross-border comparisons are involved. Indeed, this desire to eliminate accounting distortions sometimes leads to investors focusing on top-line revenue through price/sales or Enterprise Value/sales multiples. However, this often does not seem to work very well for technology and Internet stocks, partly due to accounting problems.

In most industries, the revenue line is regarded as being relatively clean of accounting distortions. This is not so for technology and Internet stocks, since the recognition of revenue is one of the most problematic areas and, because of rapid growth, this can lead to vastly different reported revenue. Given the importance of sales multiples in technology and Internet valuation, these accounting issues assume great importance.

The accounting issues for Internet and technology stocks set out in Appendix 6.1 generally relate to the classification and timing of revenues and expenses. Implementation of the regulatory guidance on gross versus net revenues, barter transactions, sales incentives and upfront fees would generally result in lower revenues (or delayed revenues in the case of upfront fees).

Appendix 6.1

Key accounting issues in Internet and technology stock valuations

Issues	*Description*	*US FASB regulatory guidance*
Gross or net revenues	Some Internet companies act as intermediaries in selling the products or services of third parties, and do not bear the risks and rewards of ownership. For example, the risk of product returns or bad credit sales is borne by the third parties. A common practice among Internet companies is to recognize as revenues the entire amount billed to customers (gross revenues) rather than the amount of fees or commission (net revenues) that these companies ultimately get to keep. Extract from the annual report of Cheap Tickets for the period ended 31 December 1999: *With respect to published fares, Cheap Tickets records as revenue in its statement of operations only the commissions earned by Cheap Tickets on the sale of such fares. Gross bookings represent the retail value of the sales of published fares. With respect to non-published fares, revenues as reported in Cheap Tickets' statement of operations revenues are equivalent to gross bookings, which are the retail value of such fares. Ideally, this should only be presented as net*	If a company acts as an intermediary in a transaction, and does not bear the risks and rewards of ownership, then the company should report net revenues

Issues	*Description*	*US FASB regulatory guidance*
Advertising barter transactions	A popular type of transaction among Internet companies. Typically, two Internet companies exchange the right to place advertisements on each other's website. Both companies recognize the revenues and costs associated with this transaction. If the reciprocal transaction occurs in the same quarter, then the revenue and costs arising from this exchange offset one another with no impact on net income Extract from the annual report of Sportsline.com for the period ended 31 December 1999: *Barter transactions, in which the Company received advertising or other services or goods in exchange for content or advertising on its Web sites, accounted for approximately 17%, 19% and 19% of total revenue for 1999, 1998 and 1997, respectively*	Revenue and expense should be recognized only when the fair value of the barter transaction is readily determinable from an entity's history (not older than six months) of selling similar advertising to buyers unrelated to the counter party in the barter transaction
Website development costs	These include the costs of developing a company's website and services such as chat rooms, search engines and e-mail. This is a significant cost for many Internet companies Extract from the annual report of Ebay for the period ended 31 December 1999: *The company's sales and marketing expenses for both the online sales and traditional auction businesses are comprised*	A large portion of such costs should be treated as costs of software developed for internal use. SOP 98–1 states that software costs incurred in the

Issues	*Description*	*US FASB regulatory guidance*
Website development costs – continued	*primarily of compensation for the Company's sales and marketing personnel, advertising, tradeshow and other promotional costs, expenses for creative design of the Company's website and shared employee and facilities costs*	preliminary stage should be expensed, but once certain technical criteria had been met, direct costs and interest cost should be capitalized
Rebates, discounts and other sales incentives	These refer to certain sales incentives offered by Internet companies. For example, free one-month service, discount coupons offered by e-tailers, or rebates offered by ISPs in exchange for customers' agreement to subscribe to Internet service for three years. These costs are commonly included in marketing expense Extract from the annual report of Barnesandnoble.com for the period ended 31 December 1999: *B&N.com has included in marketing expenses approximately $6 million, or 3% of sales, of coupon redemptions for the full year ended 1999. No coupons were redeemed in 1998. Had B&N.com treated coupon redemptions as a reduction to sales and marketing expense in 1999, gross margin would have decreased from 21.0% to 18.6% and marketing and sales expense as a percentage of sales would have decreased from 55.1% to 53.7%.*	Such rebates and discounts should be considered a reduction of revenues

Issues	*Description*	*US FASB regulatory guidance*
One-time fees	Some Internet companies charge upfront fees for services to be provided over a period of time. For example, auction listing fees, fees for access to or maintenance of a website, or fees to provide advertising. Some companies reportedly recognize these fees as revenues immediately Extract from the annual report of Ebay for the period ended 31 December 1999: *The Company has reviewed Securities and Exchange Commission Staff Accounting Bulletin ('SAB') No. 101, 'Revenue Recognition in Financial Statements', and its effect on the recognition of listing and featured item fee revenue* *Although the Company believe the effect of SAB 101 on historical listing and featured item fee revenue is insignificant, eBay has adopted the provision for placement fee revenue in the first quarter of 2000. As such, listing and featured item fee revenue will be recognized rateably over the estimated period of the listing for sale*	Upfront fees should be recognized rateably over the period during which service is provided

Source: UBS Warburg Global Equity Research, *Navigating the I-Valuation Jungle*, May 2000

7

*Bubbleology, stock markets and the Internet and technology stock price collapse**

> Bubbles generally are perceptible after the event. To spot a bubble in advance requires a judgement that hundreds of thousands of informed investors have it all wrong.
> (Testimony of Fed Chairman Alan Greenspan, before the Joint Economic Committee, US Congress, 17 June 1999)

Alan Greenspan, Chairman of the US Federal Reserve, made the remark that opens this chapter well before the collapse in the price of technology stocks. However, as will be seen in this chapter, the whole issue of whether 'bubbles' in stock markets do or do not exist raises many complex issues. Greenspan's comments, particularly his earlier one in 1996 that financial markets suffered from 'irrational exuberance', opened up the debate as to whether stock markets are prone to speculative bubbles and, if so, as to how financial economists can identify them.

The Internet and technology stock price collapse

Following Greenspan's comments one can look back and see what lessons can be learned from the year 2000/2001 technology stock price crash. With the wisdom of hindsight, we can see

*The contents of this chapter are discussed in more detail in *Financial Economics* by Brian Kettell, published by Financial Times–Prentice Hall, 2001.

that the Internet bubble was, in many ways, like the other bubbles in history. However several unique factors contributed to the suddenness of this one. First, people suddenly realized that this peculiar industry in California was making a huge amount of money, which in turn powered the system. Secondly, the Internet revolution was real, and US productivity growth rapidly tripled to more than 3% per year in the latter part of the twentieth century. The US economy boomed, corporate profits grew even faster, and technology came to dominate capital spending. These developments spread to the entire technology sector, with the bubble quickly following.

But thirdly, the Internet bubble was uniquely propagated by the Internet itself. The revolution was not merely real – it was also self-advertising because it was universal. Suddenly anyone could trade stocks virtually for free – via the Internet; you could see financial reports and stock research about Internet companies – on the Internet; every company could prove that it had an Internet strategy – on the Internet; you could tell your friends about your marvellous adventures – again, all through the Internet.

Bubbles in stock markets

Movements in prices in any market which are thought to be self-fulfilling prophecies are often called 'bubbles' and proponents of the idea that bubbles in financial markets occur are usually referred to as 'bubbleologists'. Examination of the occurrence of bubbles in financial markets is often traced from John Maynard Keynes, who observed in 1936:

> Professional investment may be likened to those newspaper competitions in which the competitors have to pick out the six prettiest faces from a hundred photographs, the prize being awarded to the competitor whose choice most nearly corresponds to the average preferences of the competitors as a whole; so that each competitor has to pick, not those faces which he himself finds prettiest, but those which he thinks likeliest to catch the fancy of other competitors, all of whom are looking at the problem from the same point of view. It is not a case of choosing those which, to the best of one's judgement, are really the prettiest, nor even those which average opinion genuinely thinks the prettiest. We have reached the third degree where we devote our intelligences to anticipating what average opinion expects the average opinion to be. And there are some, I believe, who practice the fourth, fifth and higher degrees.

The stock market, Keynes wrote, 'is a game of musical chairs, of Snap, where the winner is the one who makes his move fractionally ahead of everyone else' (1936). So the stock market, in Keynes's view, is an environment in which speculators anticipate 'what average opinion expects average opinion to be' rather than focusing on factors fundamental to the market itself, including expected future dividends etc. The implication here is that if bubbles exist in asset markets, prices will differ from their long-term fundamental values.

It is only recently that serious research has taken place on the existence of bubbles. Economic theory, until recently, placed essentially no restrictions on how agents formed expectations of future prices. Thus a folklore of bubbles grew up. These would include the tulip bubble in seventeenth century Holland, the South Sea bubble, in eighteenth century England and the increase in stock prices during the 1920s in the United States. All these events, when followed by subsequent collapses in asset values, have been labelled as bubbles. However, the widespread adoption of rational expectations, discussed below, provides a model amenable to the empirical study of bubbles.

But before we proceed further, we first of all need to describe the terminology used by 'bubbleologists'.

Bubble terminology

Following the technical language of economics, a 'bubble' is any deviation from 'fundamental values', whether up or down. Fundamental values are a concept easier to define in theory than in practice. This concept refers to the prices stocks ought to sell for based on business's real economic value, apart from speculation. The assumption is that stock prices will ultimately (whenever that is) return to their fundamental values, however much extraneous factors may be influencing them at any one moment.

A bubble is an upward price movement over an extended range which then implodes. An extended negative bubble is a crash. 'Noise' refers to small price variations about fundamental values. So a bubble is a situation in which the price of an asset differs from its fundamental market value. With a rational bubble, as discussed below, investors can have rational expectations that a bubble is occurring because the asset price is above its fundamental value but they continue to hold the asset anyway. They might do this because they believe that someone else will buy the asset for a higher price in the future. In a rational bubble, asset prices can therefore deviate from their fundamental value for a long time because the bursting of the bubble cannot be predicted and so there are no unexploited profit opportunities.

When a stock is experiencing a rational bubble, the rate of return on an asset is higher than normal in order to compensate the holder for the possibility that the asset's price might suddenly collapse to its fundamental value. With what is known as a rational bubble, the asset holder is willing to accept a lower rate of return since the asset's price might suddenly jump up to its fundamental value.

The role of expectations in analysing stock market bubbles

Beliefs about the future are an important determinant of behaviour today. Important disagreements between differing views of how expectations are formed lie at the centre of the discussion as to the existence or otherwise of 'bubbles'. Different views about expectations can be usefully broken down into three groups: exogenous expectations, extrapolative expectations and rational expectations.

Exogenous expectations

Some economists remain almost completely agnostic on the vital question of how expectations are formed. When analysing the behaviour of the economy they simply treat expectations as exogenous, or given. Expectations are one of the inputs to the analysis. The analysis can display the consequences of a change in expectations. For example, an increase in expected future profits might increase the stock price of a firm. But this assumption of exogenous expectations means that the analysis does not investigate the *cause* of the change in expectations. In particular, it is unrelated to other parts of the analysis. With given expectations, there is no automatic feedback from rising output to expectations of higher profits in the future.

Thus, at best, economists using exogenous expectations in their analysis give an incomplete account of how the economy works. At worst they completely neglect some inevitable feedback from the variables they are analysing to the expectations that were an input to the analysis. On the other hand, since modelling expectations remains a contentious issue, proponents of this approach might argue that the various types of possible feedback on expectations can be explored in an ad hoc manner.

Extrapolative expectations

One simple way to make expectations endogenous, or determined by what is going on elsewhere in the analysis, when discussing the existence or otherwise of bubbles, is to assume that investors forecast future profits by extrapolating the behaviour of profits in the recent past, or extrapolate past inflation in order to form expectations of inflation in the near future. Proponents of this approach suggest that it offers a simple rule of thumb and corresponds to what many investors seem to do in the real world.

Rational expectations

Suppose the rate of money growth is steadily increasing and inflation is steadily accelerating. Extrapolating past inflation rates will *persistently* underforecast future inflation. Many economists believe that it is implausible that people will continue to use a forecasting rule that makes the same mistake (underforecasting of future inflation, say) period after period. The hypothesis of rational expectations makes the opposite assumption: on average, investors guess the future correctly. They do not use forecasting systems that systematically give too low a forecast or too high a forecast. Any tendency for expectations to be systematically wrong will quickly be detected and put right.

This in no way says that everybody gets everything exactly right all the time. We live in a risky world where unforeseeable things are always happening. Witness the September 11th 2001 attacks on New York. Expectations will be fulfilled only rarely. Rational expectations theory says that people make good use of the information that is available today and do not make forecasts that are already knowably incorrect. Only genuinely unforeseeable events cause present forecasts to go wrong. Sometimes investors will underpredict and sometimes they will overpredict. But any systematic tendency to do one or other will be noticed and the basis of expectations formation will be amended until guesses are on average correct.

The existence or otherwise of 'bubbles' depends on which one of these three models of expectations one applies. The adoption of the rational expectations assumption has clarified considerably the nature of price bubbles. With rational expectations a researcher can specify a model of bubbles, which is then testable. If the expected rate of market price change influences the current market price, the researcher has a model to work with. This is not to say that this is straightforward. There is an indeterminacy in the model, as a researcher is in fact faced with something to explain, the market equilibrium price, with two variables, the market price and the expected rate of market price change, both of which are interrelated within the economic system.

A bubble can arise when the actual market price depends positively on its own expected rate of change, as normally occurs in asset markets. Since agents forming rational expectations do not make systematic prediction errors, the positive relationship between price and its expected rate of change implies a similar relationship between price and its actual rate of change. In such conditions, the arbitrary, self-fulfilling expectation of price changes may drive actual price changes independently of *market fundamentals*. This situation is referred to as *a price bubble*.

An explicit definition of market fundamentals depends on a particular model's structure; indeed, the very notion of a bubble can make no sense in the absence of a precise model detailing a market's operation. Without such a model, it is impossible both to define market fundamentals and then to isolate them from the presence, or otherwise, of a bubble.

Bubbles and the efficient market hypothesis (EMH)

Efficient market theory applies the theory of rational expectations to the pricing of securities. The EMH comes in three versions: the weak form, the semi-strong form and the strong form.

- The **weak form** of the EMH asserts that prices fully reflect the information contained in the historical sequence of prices.

Thus, investors cannot devise an investment strategy to yield abnormal profits on the basis of an analysis of past price patterns.

- The **semi-strong** form of the EMH asserts that current stock prices reflect not only historical price information but also all publicly available information relevant to a company's securities. If markets are efficient in this sense, then an analysis of balance sheets, income statements, announcements of dividend changes or stock splits or any other public information about a company (the technique of fundamental analysis we discussed in Chapter 4) will not yield abnormal economic profits.
- The **strong form** of the EMH asserts that all information that is *known* to any market participant about a company is fully reflected in market prices. Hence, not even those with privileged inside information can make use of it to secure superior investment results. There is perfect revelation of all private information in market prices.

The theory of rational expectations states that expectations will not differ from optimal forecasts (the best guesses of the future) using all available information. Rational expectations theory makes sense because it is costly for people not to have the best forecast of the future. The theory has two important implications: (a) if there is a change in the way a variable moves, there will be a change in the way expectations of this variable are formed, too, and (b) the forecast errors of expectations are unpredictable.

The lessons of the EMH are that bubbles are impossible because markets are 'efficient', that is, prices reflect all available information about an asset. Adherents of EMH, believing that stocks are always correctly priced, tend to deny a connection between excessive speculation and subsequent economic crises. However, the necessary assumptions underlying the EMH must be simultaneously held. Thus it is necessary to examine the extent to which they do hold.

The stock market crash of 19 October 1987 should make one question the validity of efficient markets and rational expectations. EMH critics do not believe that a rational market place could have produced such a massive swing in share prices. To what degree should the stock market crash make us doubt the validity of rational expectations and efficient markets theory?

Nothing in rational expectations theory rules out large one-day changes in stock prices. A large change in stock prices can result from new information that produces a dramatic change in optimal forecasts of the future valuation of firms. Some financial economists have pointed out that there are many possible explanations for why rational explanations of the future value of firms dropped dramatically on 19 October 1987: moves in Congress to restrict corporate take-overs, the disappointing performance of the trade deficit, congressional failure to reduce the budget deficit substantially, increased fears of inflation, the decline of the dollar, and increased fears of financial stress in the banking industry. Other financial economists doubt whether these explanations are enough to explain the stock market drop because none of these market fundamentals seems important enough.

One lesson from the 1987 Black Monday stock market crash appears to be that factors other than market fundamentals may have had an effect on stock prices. The crash of 1987 has therefore convinced many financial economists that the stronger version of efficient markets theory, which states that asset prices reflect the true fundamental (intrinsic) value of securities, is incorrect. They attribute a large role in the determination of stock prices to market psychology and to the institutional nature of the market place. However, nothing in this view contradicts the basic reasoning behind rational expectations of efficient markets theory – that market participants eliminate unexploited profit opportunities. Even though stock market prices may not always solely reflect market fundamentals, this does not mean that rational expectations do not hold. As long as the stock market crash was unpredictable, the basic lessons of the theory of rational expectations hold.

Rational bubbles

Keynes (1936), as mentioned earlier, is noted for his observation that stock prices may not be governed by an objective view of 'fundamentals' but by what 'average opinion expects average opinion to be'. His analogy for the forecasting of stock prices was that of trying to forecast the winner of a beauty contest. Objective beauty is not necessarily the issue; what is important is how one thinks the other judges' perception of beauty will be reflected in their voting patterns.

Rational bubbles arise because of the indeterminate aspect of solutions to rational expectations models, which for stocks is implicitly reflected in what is known as the Euler equation for stock prices. This states that the price you are prepared to pay today for a stock depends on the price you think you can obtain at some point in the future. The standard form of the Euler equation determines a sequence of prices but does not 'pin down' a unique price level. However, in general the Euler equation does not rule out the possibility that the price you pay today may contain an explosive bubble.

While one can certainly try to explain prolonged rises or falls in stock prices as due to some kind of irrational behaviour such as 'herding', or 'market psychology', nevertheless recent work emphasizes that such sharp movements or 'bubbles' may be consistent with the assumption of rational behaviour. Even if traders are perfectly rational, the actual stock price may contain a 'bubble element' and therefore there can be a divergence between the stock price and its fundamental value.

So proponents of rational bubbles attempt to demonstrate how the market prices of stocks may deviate, possibly substantially from their fundamental values even when agents are homogenous, rational and the market is informationally efficient. To do this they must show that the market price may equal its fundamental value plus a 'bubble term' and yet the stock will be willingly held by rational agents and no supernormal profits can be made.

It must be stressed that firm conclusions about the existence or otherwise of speculative bubbles are difficult to establish. There are severe econometric difficulties in testing for rational bubbles. Such tests critically depend on the correct specification for asset returns. Rejection of the no-bubble hypothesis may well be due to mis-specifying the underlying model of the fundamentals.

Some famous bubbles in history

It is instructive to examine some of the most famous bubbles in history that have ended in speculative collapse. The most analysed have been the Tulipmania in Holland in 1636, the South Sea Bubble in England, 1711–1720, and the 1929 Wall Street Crash. Recent empirical work, particularly on Tulipmania and the South Sea Bubble (Garber, 1989a, b, 1990), has cast doubts on whether these were bubbles. But it is still useful to paint the picture.

Tulipmania

The first account of tulips in Europe is from 1559, when a collector of exotic flora, Councillor Hewart, received a consignment of tulip bulbs from a friend in Constantinople, which he planted in his garden in Augsberg, Germany. His tulips drew a good deal of attention and in following years this flower became more and more popular among the upper classes, particularly in Germany and Holland, where it became the custom to order bulbs at exorbitant prices directly from Constantinople. Up to 1634 this custom became increasingly common, and from that year affluent society in Holland considered the lack of a tulip collection to be proof of poor taste.

Year by year tulip bulb prices rose, finally reaching astronomic heights. According to original accounts of the peak of the tulip mania, in one deal the following price was paid for one tulip bulb of the rare 'Semper Augustus' variety: 4600 florins, a new carriage, two grey mares and a complete bridle and harness. In today's money 4600 florins is $US1.5 million! One single bulb of another rare variety, 'Viceroy', was sold for 24 carriage loads of grain, eight fat hogs, four cows, four barrels of ale, 1000 pounds of butter and a few tons of cheese.

In early 1636 demand for tulip bulbs had risen so drastically that people started to trade them on exchanges in a number of Dutch towns. Tulips were no longer bought only by well-to-do collectors but also by investors and speculators. At the smallest price drop they bought up, to sell later at a profit. To facilitate trading on margin, tulip options were introduced, requiring a margin deposit of only 10–20%. Ordinary people in all business sectors started to sell off assets to invest in this attractive market.

The Dutch tulip boom also drew attention from abroad and capital began to stream into the market. This capital forced up prices for land, property and luxury goods, as well as tulips, to new record heights. Fortunes grew and a growing *nouveau riche* group was added to the old upper classes. This new affluent class had earned its money from, and reinvested in, tulip bulbs. The story is told of a brewer in Utrecht who went so far as to exchange his brewery for three valuable tulip bulbs.

In September and October of 1636 market psychology began to alter and doubts began to emerge. How could one be *sure* that three tulip bulbs were worth as much as a brewery? Suppressed mirth began to be heard. Who said a tulip bulb was worth anything at all? The market was seized by panic and prices began to plummet.

Many of the *nouveau riche* had to face the fact that they owned a fortune consisting only of tulip bulbs which nobody wanted, less broker cash loans which they could not repay. The government tried to find a compromise by declaring all tulip contracts from before November 1636 as being invalid, while all subsequent contracts would be honoured at 10% of original value. But prices dropped below this 10% and the number of bankruptcies increased day by day. Tulipmania was followed by a depression in Holland from which it took the country many years to recover.

Garber (1989a, b) points out that the standard version of Tulipmania neglects discussion about what the market fundamental price of bulbs should have been. To form an expectation about the price of tulip bulbs, Garber collected data on bulb price patterns for various highly valued tulip bulbs. He found that the extremely high prices reported for rare bulbs and their rapid decline reflected normal pricing behaviour in bulb markets and cannot be interpreted as evidence of market irrationality. Garber points out that serious traders ignored the market and participants in the market had almost no wealth anyway. Garber concludes that tulip prices at the time could be explained by market fundamentals and that Tulipmania does not qualify as being a bubble. It must be stressed that his findings have been hotly disputed.

The South Sea Bubble

A second instructive example of bubbles was the speculation in England at the beginning of the eighteenth century. The Earl of Oxford founded the South Sea Company, financed by a number of the merchants of that time, in 1711. The company's full name was 'The Governor and Company of the Merchants of Great Britain to the South Seas and other parts of America for the encouragement of the Fishing'. The company acquired almost 10 million pounds of the British national debt, against a guaranteed annuity of 6%, and the monopoly of all trading with Latin America.

A short time after the company's founding, rumours arose of incredible profits to be had from trading in the South Seas, where it was said English goods could be bartered for gold and silver from the 'inexhaustible' mines of Peru and Mexico. In fact, the Spanish colonial power allowed only one English ship to call per year, for which it charged one-quarter of all profits and 5% of turnover. On the stock exchange the South Sea stock led a quiet existence, the price often moving only two or three points over a month.

In 1717 the King recommended that the national debt be 'privatized' once more. The country's two large financial institutions, the Bank of England and the South Sea Company, each submitted a proposed solution and, after heated parliamentary debate, it was resolved to allow the South Sea Company to acquire a further debt liability at an interest rate of 5% per year.

But in 1719 an event took place in France, which was to be of great significance for the English company. A well-to-do man named John Law had founded the 'Compagnie d'Occident' in Paris, to trade with and colonize the American state of Mississippi. By a series of manipulations, John Law succeeded in starting a massive wave of speculation in this company's stock, the price rising from 466 francs on 9 August to 1705 francs four months later. Buyers were French and foreigners alike, which caused the British Ambassador to request His Majesty's Government to do something to stop the massive flow of English capital to the 'Mississippi Bubble' on the French stock exchange. The Mississippi Bubble collapsed on 2 December 1719, and in the ensuing crash investors, seeking profitable opportunities, moved their funds from France to England.

Further privatization of the national debt provided an interesting opportunity for the principal stockholders in the South Sea Company, who now offered to take over the entire debt of the English state. On 22 January 1720 the House of Commons appointed a committee to consider the proposal. Despite many warnings, on 2 February the decision was taken to submit a bill to Parliament. Investors were delighted at this prospect of further capitalization of the company and over a few days the share price rose to £176, supported by an inflow of funds from France. During further readings of the bill new rumours started to circulate on the unbelievable profits that could be made and stocks rose further to a price of £317.

Even at this price the company's original founders and co-directors could reap a capital gain that was enormous by the standards of that time, and in a still virtually inactive company. This whetted their appetites for more, and new positive rumours were circulated; on 12 April fresh stock was subscribed to for one million pounds at a price of £300. The issue was subscribed twice over and a few days later stock was traded at £340. The company then declared that a 10% dividend would be paid on all new and old stock and a further new subscription was invited for one million pounds at a price of £400. This was also over-subscribed. The company was still almost totally inactive.

Many other companies jumped on this speculative bandwagon, issuing their own shares. However on 11 June 1720 the King proclaimed a number of these companies to be 'public nuisances' and trading in their stocks was prohibited on penalty of a fine. Mackay (1841) describes a list of 104 companies that were banned.

Despite the government's endeavours, new bubbles appeared every day and the speculation fever continued to rise. Figure 7.1 charts the progress of the South Sea Bubble. The South Sea Company was traded at a price of £550 on 28 May 1720. From this already impressive level, during June the price rose above £700. In this period price movements were extremely nervous, with great periodic shifts. On a single day, 3 June, the price dropped before noon to £650, to rise again in the afternoon to

Figure 7.1 The South Sea Bubble, 1719–1720. *Source: The Psychology of Finance*, L. Trede. Reproduced with permission of John Wiley & Sons

£750. Many large investors used the high prices to take profits that were reinvested in anything from land and commodities to real estate and other stocks. However, others re-bought the South Sea Company's stock, one of them the physicist Isaac Newton. During the stock's early rises he had sold all his South Sea stock, cashing a profit of £7000. In midsummer he bought again, a transaction that would come to cost him £20 000.

At the beginning of June, South Sea stock rose again and for a short enchanted moment, on 24 June 1720, the security was traded at £1050. As only few were aware, the time was running out for investors. Those in the know were the company's original founders and its Board Chairman, who had used the earlier high prices to get rid of their own stock. At the beginning of August this ominous fact began to leak to the general public and the stock price began to fall slowly and steadily.

On 31 August the South Sea Company management announced that an annual dividend of 50% would be paid for the following 12 years. This would have completely drained the company of cash and the news did not stop the investors' increasing unease. On 1 September the stock continued to fall and when it reached £725 two days later, panic broke out. The security went through the floor over the rest of the month and when the company's bank was declared bankrupt on 24 September the fall accelerated. On the last day of the month the share could be bought at a price of £150. In only three months it had fallen by 85%.

The company was finally dissolved in 1855 and its stock converted to bonds. In its 140 years of existence the company never succeeded in trading in the South Seas on any noteworthy scale.

The Wall Street Crash of 1929

The 1929 Wall Street Crash was the conclusion of one of history's largest episodes of mad speculation, but, as the recent technology stocks crash demonstrates, almost certainly not the last. For a number of years up to 1924 the American Dow Jones Industrial Index fluctuated within a relatively narrow price interval, with strong selling pressure whenever it reached 110. From 1921, when the stock market was very depressed, to 1928, industrial output rose by 4% annually and by 15% from 1928 to 1929. Inflation was low and new industries sprouted forth everywhere.

This rising optimism, combined with easy access to cheap money, stimulated stock investors and after a temporary reversal in 1926 almost no month passed without a rise in stocks creating a new generation of rich investors. Investment trusts increased in number as stock investments rose in popularity. From around 40 companies before 1921 the number rose to 160 at the beginning of 1927 and 300 at the end of the same year. From the beginning of 1927 to the autumn of 1929 the total assets of investment trusts increased more than ten-fold and there was almost unlimited confidence in these companies.

On 24 October 1929, trading reached 12 million stocks. Nervousness, however, had set in and a panic was evident. As the situation was clearly getting out of hand, on 25 October President Hoover made the following statement: 'The fundamental business of the country, that is, production and distribution of commodities, is on a sound and prosperous basis.'

Hoover's declaration had the same reassuring effect as a ship's captain announcing that the ship was *not* sinking. Panic grew and in the next few days prices continued to fall. This

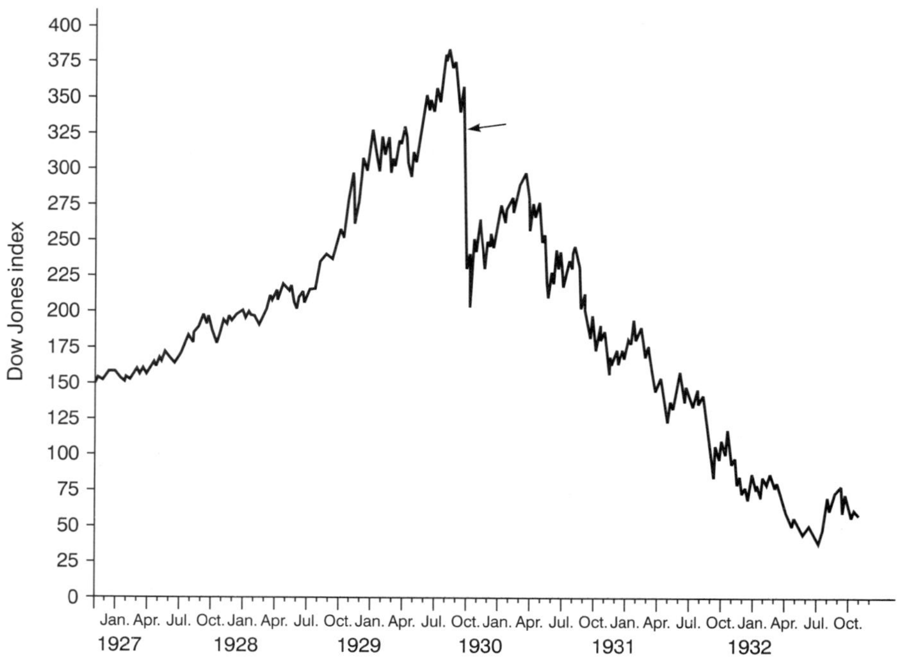

Figure 7.2 The stock market crash of 1929 (arrow denotes Hoover's declaration that 'the fundamental business of this country . . . is on a sound and prosperous basis'). *Source: The Psychology of Finance*, L. Trede. Reproduced with permission of John Wiley & Sons

culminated on 29 October when, in a wave of enforced sales, 16 million stocks were realized at any price going. The story goes that a messenger at the stock exchange got the idea of bidding a dollar per share for a lot without buyers – and got his deal. Prices did not start to stabilize until the index reached 224, on 13 November, as shown in Figure 7.2.

In 1930 prices started to fall once more, continuing to a bottom of 58 on 8 July 1932.

Speculative bubbles theory

According to speculative bubbles theory, stock market investment is more about inflating and bursting speculative bubbles than about investors making rational long-term forecasts. Under speculative bubbles theory an asset's price can be bid up above its intrinsic value because some market participants believe that others will be willing to pay still more for it tomorrow. For a while this belief is self-sustaining and the market booms, but eventually participants lose faith that prices can rise any further and the market crashes.

This theory, also named 'Castles in the Air' theory by Malkiel (2000), based upon the thrills of making a killing by selling castles in the air, has long antecedents, only some of which have been discussed above. McKay (1841) provides an impressive list of South Sea Bubble emulators. Companies were created, for example, to design a wheel of perpetual motion, to extract silver from lead and to build ships to defeat pirates, shares of all of which had the characteristics of speculative bubbles.

The prize, however, must go to the unknown soul who started 'a company for carrying on an undertaking of great advantage but nobody knew what it was'. The prospectus promised unheard of rewards. At nine o'clock in the morning, when the subscription opened, crowds of people from all walks of life practically beat down the door in an effort to subscribe. Within five hours some £4000 pounds sterling had been handed over. The promoter promptly closed up shop and was never heard of again!

Rational speculative bubbles

Hardouvelis (1988) has argued that speculative bubbles may be triggered by an extraneous event that is unrelated to fundamental economic conditions; one group of investors buys with the expectation of a large capital gain, and others follow suit, without paying proper attention to economic factors such as future dividends or interest rates. If such behaviour persists, it may feed on itself as consecutive waves of buying increase prices. Speculative bubbles may subsequently burst very suddenly. An overvalued market is fragile and a relatively unimportant piece of 'bad' news may easily create pessimism and set off a selling wave.

The traditional method of searching for market overvaluation or speculative bubbles counts the number of unusually high returns during the suspected bubble period and assesses the likelihood that the total number of these high returns could have arisen from chance (Blanchard and Watson, 1982). An unusually high return (or a positive 'abnormal' return) is a return higher than the risk-free rate plus the usual risk premium necessary to compensate risk-averse shareholders for the uncertainty associated with their security returns. In the absence of a speculative bubble, a very large number of unusually high returns would normally occur by chance only with a small probability. Hence, a large number of unusually high returns constitutes evidence consistent with the presence of speculative bubbles.

Unfortunately, although simple, the traditional test has low statistical power to detect speculative bubbles; share prices are very volatile and their swings generate both large positive and large negative returns. The latter tend to mask any existing bubble evidence.

In order to construct a more powerful test for bubbles, it is necessary to formulate a more precise economic account of the development of the bubble. One can imagine many different scenarios of market overvaluation, but analysis restricts the possible scenarios to those in which investors know that the market is overvalued yet show no special desire to liquidate their positions and continue to buy or sell as they would in the absence of bubbles. This is a realistic working assumption for the period before October 1987. Robert Shiller (1987) provides survey evidence indicating that, before October 1987, 71.7% cent of individual investors and 84.3% of institutional investors thought that the market was overvalued at the time. Schiller argued that the crash was generated by what he calls a 'feedback loop'. After a first price decline investors sold, not based on fundamentals, but because they were worried about what was going on and about market irrationality. The drop did not stop until enough people started to have opposing feelings.

Explaining why investors did not get out of an overvalued market is more difficult. One could argue that the presence of highly liquid futures markets and associated trading strategies such as portfolio insurance led investors to the false belief that they could enjoy large positive returns in an upward market yet still avoid suffering a large loss if the market took a big plunge.

An alternative explanation, outlined above by Hardouvelis (1988), is one that does not depend on some sort of collective irrationality. Within the economics literature this is known as the 'rational speculative bubble hypothesis'.

The 'bubble premium'

In the case of a rational speculative bubble, investors know that the bubble may crash and that they will not be able to get out once the crash starts, but they remain in the market because they believe – for whatever reason – that there is a good probability that the bubble will continue to grow, bringing them large positive returns. These returns are expected to be higher than the risk-free rate plus the usual risk premium in the absence of bubbles, and large enough to compensate them exactly for the probability of the bubble crash and a large one-time negative return. Hence, it is rational for investors to stay in the market. The expected extra return when no bubble crash occurs can be called the 'bubble premium'. The theory implies that the bubble premium is not only positive, but also increases during the lifetime of the bubble. The time trend in the bubble premium derives from the explosive nature of the bubble component of the stock price. As time goes on, the bubble component of the stock price grows larger and larger relative to the fundamental components. This growth implies that with the passage of time, the expected drop in the stock price in the case of a bubble crash grows larger too, necessitating a larger and larger bubble premium.

Was the Internet and technology stock price crash a predictable bubble?

Historically the following can be said about bubbles:

(1) Bubbles grow at an exponentially increasing growth rate. The growth rate is highest just before the crash.
(2) Bubbles always grow larger and last longer than anyone expects.
(3) The crash is a total surprise, coming just when it seems everyone has accepted that it will continue forever. In fact this is what precipitates the crash. There are no new fools left to bid up the price, everybody is already in.
(4) The base of the bubble narrows at the top. At first the best technology stocks (tulip stocks etc.) go up, then all technology stocks go up. At the end a select few technology stocks go ballistic, then all technology stocks crash.
(5) Most fortunes made during the bubble are lost. The enduring fortunes are made by those able to avoid the crash and buy cheap real assets after the crash, a process which certainly took place in the early years of this new century.

The wisdom of hindsight is a wonderful thing. In principle, few of the participants in the recent Internet and technology stock price collapse would place their hand over their hearts and deny they would do the same thing again. And so it goes on!

8

How should we value Internet and technology stocks? I: Key methodology

The key to successful valuation of Internet and technology stocks is to apply techniques that can predict *future* cash flows, given that *current* profits and cash flow may give little help in forecasting the future prospects for a company.

Key issues in valuation of technology stocks

Some of the key areas complicating the valuation of these stocks were discussed in Chapter 6.

The valuation issues revolve around:

- Profit and cash flow earned today is, for many technology companies, not a good predictor of future cash flow generating ability and hence of value. This places greater emphasis on other less reliable performance measures and upon speculative forecasts.
- Most of the value inherent in these stocks is further out. Near-term cash flows are small or negative and the ultimate hoped-for cash flow and value creation may not materialize for many years.

- A greater proportion of the value of technology companies is inherent in growth opportunities. Value is not a function of what these companies are doing today but is dependent upon what they could be doing in the future.
- All the issues revolving around the valuation of intangible assets, as discussed in Chapter 6, complicate the valuation exercise.

Value drivers: where is the fundamental value in technology companies?

The fundamental value of any investment is a function of the future cash flows it is expected to generate, the so-called 'value drivers', and the return investors require from the investments. Each of these two components is itself a function of several different elements.

The value drivers for technology companies are highlighted in Table 8.1.

The value of growth stocks is highly sensitive to changes in these value drivers. It is therefore not surprising that prices of these stocks are so volatile.

Valuation methods for technology stocks

In this chapter we set out some of the principles and techniques that need to be applied in valuing Internet and technology stocks. These ideas draw on the methodology proposed by investment bank UBS Warburg. We concentrate here on three techniques:

(1) **Current priced multiples of sales revenues:** This is a useful relative comparison for all companies where there is a lack of current profits suitable as a basis for valuation. Interpretation must be done with care and particularly with reference to anticipated future margins.

(2) **Forward priced multiples of forecast revenues or profits:** This is a method of deriving a target absolute value based upon the anticipated position of the company at maturity. This approach can also be used for any company where there are no current profits on which to base value. Using

Table 8.1 Value drivers for technology companies

Value driver	*Characteristics*
Growth	The speed that a technology company can grow its business is critical. Investors have an ever-shortening time horizon over which they will tolerate the time between the listing of a technology company and the period it can show critical mass. Investors look to revenue, profit and cash flow growth
Quality of growth	A higher quality of growth should attract higher valuation levels. Technology companies that can grow with 'value addition' – and that can sustain that value addition – will attract the highest valuation levels over time
Business margins	There are many business models for listed technology companies. Companies that have businesses that deliver higher margins will be more highly valued than companies generating larger revenues yet at lower margins
Customers and marketing	The likely behaviour of the customers of technology businesses is not well understood. How loyal will they be? At what level will marketing spend stabilize? How will brand name, first mover and size affect customers? These are key components of both growth and margins
Capital intensity	In the nascent stage of development technology companies require a large amount of cash. This cash is primarily being utilized to market their businesses and to develop new applications and services. The greater the amount of cash required to generate business development the greater the degree of risk the investment holds
Premium return on investment	Technology stocks price the value of the growth and profitability of the ongoing business as part of the total valuation. Investors allocate a substantial amount of value to the ability of technology companies to generate additional growth from the success that they will experience. The value of the ability to continue to find new investment opportunities which can deliver high levels of return, which are measured by their excess over the cost of capital, is the 'premium' return on investment

Source: Adapted from UBS Warburg

the forward priced multiple method forces investors to focus on what the company will look like at maturity. This method is consistent with both discounted cash flow methodology and is also effectively based on a profit multiple. Where a company is in a more mature stage then the emphasis should shift from one focused on revenues to one focused more on profit multiples.

(3) **Discounted cash flow:** Discounted cash flow (DCF) can be difficult to apply at the very early stage of a developing, high-growth company's life and would perhaps carry less weight in analysis, as we discussed in Chapter 6. But as short-term forecasts start to show cash flows, which are more reliable as a basis for longer-term projections, this method should be given more attention.

We now turn to the three techniques in more detail.

Current priced multiple of sales revenue

As stressed throughout this book, valuation multiples should ideally be based on profits rather than revenues, since it is profit and the cash flow derived from it which creates value. In the absence of current profits for most technology-based companies, investors were forced to look to other measures. Sales revenue is the next best thing to profit, although clearly it has drawbacks. However, one may expect those stocks where the future margins will be highest to sell for higher sales multiples, assuming all other value drivers are equal.

Margin-adjusted sales multiples

In making comparisons of stocks using revenue multiples it is vital that only those with similar margin potential should be included. Profit-based multiples would be more comparable since these are not directly affected by margin. However, revenue multiples for different business models can be compared if they are 'potential margin adjusted'. A margin-adjusted sales multiple of a competitor company, adjusted for the purpose of comparison with a target company, can be calculated using the following formula:

$$\text{Margin-adjusted sales multiple} = \frac{\text{Observed sales multiple} \times \text{Target company's expected operating margin}}{\text{Comparator company's expected operating margin}}$$

Use of this adjustment introduces further subjectivity into the valuation process but it should provide a better basis for making comparisons between different business models.

Forward priced multiple valuation

This technique is essentially a combination of the simplicity of a valuation multiple with the forward-looking elements of discounted cash flow.

The mechanics of forward price multiples are relatively simple.

(1) Select a time horizon.
(2) Estimate investors' required return over that period (the cost of capital).
(3) Then deduct from the current value of the business the present value of the cash flows from the valuation date to that time horizon. The result of this is the present value of the implied exit value of the business.
(4) Finally, compound this value to obtain the actual implied exit value.
(5) This is then compared with a business activity measure for that future period to obtain the forward priced multiple.

In applying this valuation technique it is useful to apply a forward price to sales model and here we set out the steps investors need to go through.

Step One: Identify the period to maturity

The model is based upon an assumed level of revenue and deserved price to sales multiple at 'maturity'. Maturity can be freely selected but is intended to be several years hence; say, five to ten years. Maturity does not necessarily mean the stock is expected to be ex-growth. It is simply that the characteristics in terms of growth, margin and return on capital have reached the estimated long-term values which it is believed the company can achieve.

Step Two: Forecast revenues at maturity

Of course, forecasting revenues several years hence is difficult and speculative, but it is a key component in valuing technology stocks. This technique uses two approaches to forecasting revenue:

- analysts' short-term forecasts plus an estimated medium-term growth rate to maturity, and
- a total market size and estimated company market share (at maturity) approach.

Step Three: Estimate forward price to sales multiple

A key element in determining the forward price to sales multiple is the estimated net margin. Again, this is a difficult driver to forecast but is also an essential component of the valuation about which a view must be taken. The margin must be post-tax and a realistic estimate of the long-term profit opportunities for the business. Key factors to consider in this are the nature of the products and services, the quality of business model, barriers to entry, brand name, etc.

Step Four: Apply the investor's required rate of return

This is likely to be relatively high for most technology stocks since this reflects the required return of investors over the period where there is greatest uncertainty.

Discounted cash flow (DCF)

As we discussed in Chapter 6, there are two main problems with the standard DCF approach.

First, *the limited explicit forecast period fails to capture all the high growth*. Many growth stocks were priced on the assumption that a premium growth rate can be maintained for longer than, say, a five-year explicit forecast. This means that the growth rate in the terminal value calculation has to be a quasi-average of continuing high growth for a limited period plus the real long-term growth; in other words, a number almost impossible to estimate.

Second, *the final-year forecast cash flow is a poor basis for measuring terminal value*. In traditional two-stage DCF models, typically 70–80% of value is inherent in the terminal value calculation, whereas for growth stocks this is usually well over 90%. For technology stocks it is not uncommon for *more than 100%* of value to be in the terminal value (the cash flows in the explicit forecast period are negative).

Since the terminal value is based upon the cash flow in the final year of the explicit forecast period, this becomes the critical figure in the valuation. However, because a technology company has probably not reached maturity at this time the basis of this cash flow forecast – the margins, capital expenditure, etc. – are not appropriate.

UBS Warburg have developed a three-stage DCF model for use with all stocks, but it is particularly useful for growth stocks where the growth period is longer than that taken by the normal analyst's explicit forecast period. There are two main differences between this model and the traditional model:

(1) a medium-term growth period, and
(2) a terminal value based upon value drivers (an approach similar to the one used in the target multiple model described later).

These are discussed below.

The medium term growth rate

Much of the value inherent in growth stocks is dependent upon:

- the period over which a premium growth rate can be maintained;
- the rate of that growth; and
- the cost in terms of additional investment.

This model looks at the current price and calculates the implied medium term growth rate. This can then be compared with the likelihood of that growth continuing.

The terminal value

This can be calculated using a value driver approach to produce a target exit multiple from inputs of forecast long-term growth, cost of capital and, importantly, the incremental return on capital.

The use of multiples as valuation techniques: arguments for and against

The use of multiples (sales revenue, forecast revenue, profits etc.) are not a measure of the absolute value of a company. They are useful for investors seeking a portfolio approach to technology stock investment. Multiples are criticized because they do not explicitly take account of the drivers of stock value, required return, growth, return on capital etc. The benefit of a multiple is that it is in some way correlated to future cash flows, profits and hence stock value. Of course this link is complex, but remember that the nature of technology companies makes traditional analysis largely inapplicable.

Clearly, current profitability is correlated with future cash flow, but what if there are no current profits? One is forced to look for other proxies, such as revenue or even subscribers. The further away from cash flow these are, the less useful they must be in valuation. In order to be effective, a valuation multiple used for technology securities needs to be forward-looking and effective in linking critical success factors in technology companies' business models to value. In Table 8.2 we consider many of these multiples in more detail and evaluate their likely predictive power.

Web metrics

Web traffic measures became standard Internet company performance benchmarks that were commonly reported in the business press, voluntarily disclosed by companies at the time of their earnings announcements, and frequently mentioned as valuation parameters in analysts' reports.

Three key dimensions of traffic generating performance are: the attraction of new visitors (or 'eyeballs') to a website; the retention of visitors at the site, conditional on having got them to the site for a visit; and the ability to generate repeat visits from surfers who have been attracted to the site in the past. These three dimensions of web traffic performance are commonly referred to as 'reach', 'stickiness' and 'customer loyalty', respectively.

Reach

Reach is generally defined as the number of unique individuals who visit a site, stated as a percentage of the (active or total) web surfing population. Reach is the web metric that was most frequently cited in the business press. As a performance measure, reach provides an indication of the scale of the web property's visitor base, which is a measure of how successful the company has been at attracting web surfers to their site.

Table 8.2 Relative merits of valuation statistics used for Internet and technology companies

Valuation statistics	*Assessment*
Web hits/ eyeballs/page-views	The 'web hit' measure is, of course, open to abuse and can be seriously misleading. There is no distinction between a casual surfer and a serious customer. Pages can be downloaded by one person any number of times. Even the term 'hit' has multiple interpretations from providers of statistics ***Predictive power of statistic for future cash flow: POOR***
Number of press releases	Press releases attract attention, sometimes a lot of attention, which certainly will have short-term effects on the daily (if not hourly) price of stock. However, no number of press releases will have any effect on the future cash flow of a business – Internet or otherwise. Cynics would say that as management sees the price of its stock languish they rush to their press agent to generate enthusiasm in their stock price. In instances where a press release is related to the announcement of a new business or refinement of a current business that adds to the future projected cash flow, that will add value ***Predictive power of statistic for future cash flow: POOR***
Number of external links	A questionable measure of future performance. The premise of 'build it and they will come' must hold true for this to be a value-added statistic. Links (banner ads or hot-links on other web pages sending a user to the web page of the listed Internet company) are of limited use in forecasting future revenue or cash flow streams. This business measure is more quantifiable than hits and less open in its interpretation. As with many valuation models, it is important to look at the value drivers. In this case it would not only be the quantity of external links, but the effectiveness (perhaps by measuring the links to the sales ratios) of this measure ***Predictive power of statistic for future cash flow: POOR***
Number of sales transactions	Once a customer has come to the virtual register and paid up for goods or services, there is a reasonable and predictable path towards future cash flow and profitability. This methodology is also susceptible to the churn rate factor discussed above with regard to subscribers as a value basis. In addition, the margin and cash profile must be considered when making comparisons between companies globally and within a market ***Predictive power of statistic for future cash flow: MODERATE***

Table 8.2 Continued

Valuation statistics	*Assessment*
Sales or revenue	Sales must always be interpreted in the context of forecast long-term margin. However, investors should also focus on a further key factor – the accounting treatment of revenue. There are many methodologies used for revenue recognition, a position complicated by still-evolving regulation. As a result, some investors are extremely sceptical about management teams that have adopted aggressive accounting policies for sales and will therefore value the stock on a significant yet unquantifiable discount to the market. Margin and cash profile must be considered when making comparisons between companies globally and within a market ***Predictive power of statistic for future cash flow: MODERATE/GOOD***
Subscribers/ 'Dial-up subscribers'/ unique users	Intuitively, a subscriber today should lead to cash flow and value addition in the future. However, with Internet companies, with high implicit churn rates (although accounts are not normally closed, just left dormant), having a subscriber today does not guarantee a subscriber (and hence cash flow) at a later date. To keep a subscriber active, resources (new services, marketing costs, promotions, new features) have to be expended to keep him or her interested and to increase revenue. When comparing one Internet company to another with a subscriber-based valuation, great caution must be taken to ensure that the comparison is between similar business models. Comparing a company in a low margin, and potentially poor cash flow business model (e-tailing) to an Internet company with high margins and good cash flow (media and portals) will result in misleading conclusions. In the case of portals where users/customers do not have to subscribe, reliably provided data that indicates the number of unique users (i.e. individual address hits) would have similarly useful attributes ***Predictive power of statistic for future cash flow: MODERATE***
Marketing costs per user	Some would suggest that this is the best measure of future cash flows – spend today to attract a customer tomorrow and they will spend in the future. This is a questionable measure of future performance, as the premise of 'build it and they will come' must hold true for this to be a value-added statistic. This business measure is more quantifiable than many of the 'poor'

Table 8.2 Continued

Valuation statistics	*Assessment*
Marketing costs per user – continued	statistics and less open in its interpretation. As with many valuation models, the key is to look at the value drivers. In this case, it would not only be the quantity of ads or promotions, but the effectiveness of the marketing campaign. Measuring the marketing costs to sales ratios – to compare the marketing costs of one Internet company to another, looking at marketing costs per subscriber – should give some base level for comparison ***Predictive power of statistic for future cash flow: MODERATE***
EBIT: Earnings before interest and taxes	More comparable than EBITDA, where capital intensity differs, but affected by accounting policy differences for depreciation, marketing costs, research & development costs and amortization of intangibles (i.e. goodwill on acquisitions) ***Predictive power of statistic for future cash flow: GOOD***
EBITDA: Earnings before interest, taxes, depreciation and amortization	Closer to cash flow than other profit measures, but not a true cash flow. This profit measure avoids problems of different accounting policies for depreciation and marketing/research/development costs. Not an appropriate measure to use in comparison of Internet companies where capital intensity differs. Higher capital intensity results in a lower EBITDA multiple. Forecast or current EBITDA-positive businesses will have a higher likelihood of future cash flow ***Predictive power of statistic for future cash flow: GOOD***
Ferrari owned by CEO	It's true – some people really have suggested this! This is the most backward-looking business measure, with virtually no predictive power. The scenario is: CEO made a stash by listing the company, starts selling down his share options but keeps living life on the high side (sounds like exit behaviour!). Notwithstanding, with a dash of marketing flair, this type of behaviour could have the ability to push up the share performance in the very short term ***Predictive power of statistic for future cash flow: POOR***

Source: UBS Warburg

Stickiness

Website 'stickiness' generally refers to a site's ability to retain a surfer at the site once a customer has arrived there. Website 'stickiness' is a desirable quality because a 'sticky' site may be able to generate higher advertising rates from advertisers who believe that visitors are more likely to spend sufficient time at the site to read, retain, and/or otherwise be influenced by the ads that are placed there.

Customer loyalty

Customer loyalty generally refers to a website's ability to generate repeat visits from surfers who have previously visited the site. This metric is relevant because a website's ability to re-attract current visitors is expected to be an important determinant of its ability to sustain, and/or ultimately grow to, the critical mass of traffic that is necessary to attain profitability.

Web metrics are all subject to the limitations discussed earlier. In the absence of profits they were designed to provide some indicators, albeit limited ones, of potential future cash flow.

9

How should we value Internet and technology stocks? II: Ad hoc techniques and DCF revisited

A variety of what could be called ad hoc technology stock valuation methods have acquired Andy Warhol's 'fifteen minutes of fame' status at different times. These are described in this chapter. Following this, a revised version of the discounted cash flow methodology is discussed. Perhaps, as will be seen later, there is still life in the old DCF dog after all.

Harmon's Internet toolbox

In an article entitled 'Inside the Internet valuation tool chest', published on the internetnews.com website, Steve Harmon, a senior investment analyst, proposes 19 different valuation tools that can help value an Internet company's stock. Most of these valuation methods have been used by a variety of investment companies and analysts. Such companies include T. Rowe Price and Credit Suisse First Boston Corporation. All 19 measures can be seen in Table 9.1; those methods marked * are ratios that Harmon and his colleagues claim to have invented themselves.

A few of the most important tools mentioned are further discussed below.

Table 9.1 Steve Harmon's Internet valuation metrics

Market capitalization/users*	A reference for comparables and peers, an indicator of overall value per reach
Market capitalization/page views*	Indicates what a basic page view is valued at by investors
Market capitalization/ad views*	Page views generating revenue rather than sitting blank
Private market value	What private deals go for, either in merger or in bidding situation
Customer acquisition cost	What it costs to gain a new subscriber or buyer; valuable in evaluating ISPs, e-tailers, auctioneers and wholesalers, manufacturers that sell on the Web
Valuation/customer acquisition cost*	Useful when comparing peers to see which may be valued at a relative discount based on customer growth efficiencies
Enterprise value	Subtract cash and add debt to market capitalization to determine core value of the company; particularly useful in valuing potential take-over targets
Revenue multiple (or market capitalization/revenue)	Primary method for valuing Internet stocks since most firms are not earnings positive. Or if they are, the P/E is often off the charts
EBITDA cash flow multiple	The way mature media companies are valued, such as Time Warner, Disney or TCI. Usually EBITDA cash flow
Revenue per subscriber	Primary way to value ISPs such as PSINet, Earthlink, Mindspring or AOL
Lifetime value of an e-buyer*	The metric expected to be the best future method for valuing Internet companies, especially Amazon.com, CDnow, Egghead, ONSALE and e-tailers
Effective deal value	What a deal went for after factoring cash and debt or other considerations that affect the outcome of the offer
Market capitalization/POP*	A ratio for comparing ISPs that have points of presence
Market capitalization or PMV to potential market	Helpful in determining future revenue, cash flow and earnings to see if the firm is under- or overvalued to its potential
Market capitalization/total Internet users*	The value of a firm's reach globally per user

Table 9.1 Continued

Revenue/direct e-marketing*	Shows yield of campaign efficiency, useful for direct sellers such as Xoom.com
Market capitalization/ Websteader*	Useful for valuing community of free home page providers such as GeoCities, theglobe.com or Lycos' Tripod/Anglefire
Revenue/Websteader*	Measures yields on community sites; how effective Websteads generate revenue for community firms
Price/discounted earnings	Project earnings and discount back to current price; especially useful for firms with losses today
Sales/employee	How lean and mean a company operates; 1 engineer to 10 light bulbs or 10 engineers and 1 light bulb?
Revenue/bandwidth*	Useful to determine how effectively management deploys its bandwidth to generate revenue; applies to all Web firms but especially those involved in selling
Revenue/reach*	How much revenue is generated compared to the firm's percentage reach points on the Web; divide by total reach percentage points

*Ratios invented by Steve Harmon and his colleagues during the past 5 years of analysing the Internet as an investment vehicle.
Source: Harmon, 1998. © 1998 Mecklermedia: www.isdex.com

Market capitalization/total Internet users

This ratio is worked out by dividing the market capitalization of a company by its total amount of users. This gives a measure of the value of a company's reach globally.

Revenue market (market capitalization/revenue)

The second ratio is calculated by dividing the market capitalization of the company by its total revenue. This was a primary method for valuing Internet stocks since most firms' earnings were negative.

Lifetime value of an e-buyer

This is a metric that Steve Harmon thinks will be a key method of valuing Internet stocks for the near future, although he doesn't give any idea how to work out this value. It is an estimate of how much a site can earn from a visitor over the lifetime of the relationship.

Market capitalization to industry potential market share

This method can be useful in predicting future revenue, cash flow and earnings of a company. This measure can indicate whether a firm is under or overvalued to its potential.

Angel's net stock valuation calculator

Professor James J. Angel of Georgetown University, USA provides a method that calculates the stock price of an Internet company. His calculation is based on an assumption that someday an Internet company will 'grow up' and become a mature company with predictable profits and growth that then can be valued with 'traditional' pricing models. In order to use this model, forecasts need to be made on how the company will look when it has matured. Estimates must also be made on the time frame it will take for the company to mature. Figure 9.1 below indicates the variables needed to calculate the stock price of an Internet company, in this case Amazon.com. Explanations of the individual variables can be found below along with his methodology for calculating the share price.

Net stock valuation calculator variables and methodology

Probability of success

This is a number between 0.0 and 1.0. There are many unforeseen things that can go wrong with any high growth stock, and this holds true for any Internet stock too. Thus, a proper valuation should take into account the fact that probability of surviving is less than perfect.

Revenue when firm matures (millions)

This is what the investor thinks the total revenues of the firm may be when it matures into a company with predictable earnings and growth. One way to try to predict this is to look at the total size of the market in which the company competes, and try to guess what kind of market share the firm may end up with.

Profit margin

Losses in the early years of many start-up companies are to be expected. What really matters is how much profit the firm will regularly generate after it matures. What profit margin do you think is likely for the firm that you are seeking to value?

P/E ratio

The price to earnings ratio is one benchmark of a firm's value. The market basically assigns a higher P/E multiple to firms with good growth prospects.

Number of shares

One thing to watch for is the number of shares that might be issued later due to the exercise of stock options or the conversion of convertible bonds.

Discount rate

Since the payoff for investing in a growth stock is in most cases some years in the future, it is necessary to discount the future value of the stock to find out what it is worth today. A value of 10% (0.10), Angel suggests, is a reasonable guess as to the appropriate discount rate for the stock market. Various rates can be used depending on the individual investor's expectations.

Years until maturity?

How many years will it take for the stock to grow up into a firm with steady earnings and predictable growth?

Angel's valuation model

Angel's methodology is fairly simple. First, he calculates the value per share of the firm in the future after it grows up, as shown in Step 1 below.

Price in the future = Revenues in the future
* Profit margin * P/E ratio in future Step 1

Then he discounts to today's value by taking into account the probability of success, the discount rate and the number of years to maturity. This can be seen in Step 2.

Price today = Probability of success
* Price in future / (1 + discount rate)^
(No of years to maturity) Step 2

Figure 9.1 highlights a hypothetical example of valuing an Internet stock by using this method to value Amazon. Certain assumptions must be made. Suppose an Investor assumes that Amazon.com has a 75% chance of receiving $10 billion in revenues per year in five years' time, with a 5% profit margin. The investor also believes that its P/E ratio will be 25 then and the expected discount rate for the stock will be 10% per year. Assuming that the total amount of shares outstanding is 159 million, Angel calculates Amazon's share price at $37. One must then compare this valuation with the market price to decide if it is cheap or expensive.

The problem with valuation methods that result in earnings forecasts such as Angel's valuation technique is that they require estimates of revenue growth, profit margins and other important inputs. This becomes an even bigger problem for new Internet start-up companies, with little history and vague business plans. The most logical and thoughtful estimates are still big guesses. This is especially true where discounted cash flows are used as outcomes can be very sensitive to the discount rate.

Probability of success (between 0 and 1)	Revenue when firm matures (millions)	Profit margin (%)	P/E ratio	Number of shares (millions)	Discount rate	Years until maturity
0.75	10 000	5	25	159	0.1	5

Figure 9.1 Hypothetical example of valuing Amazon.com

Discounted cash flow fights back!

As discussed in Chapter 6, discounted cash flow has some limitations when it comes to valuing dot.com companies. However Copeland, Keller and Murrin, writing in their book *Valuation: Measuring and Managing the Value of Companies* (2000), put forward a revised version of DCF enabling the technique, so they claim, to be just as applicable to dot.com companies as to standard 'Old Economy' type companies.

The authors focus on the three twists required to make DCF more useful for valuing Internet companies. These are: start from a fixed point in the future and work back to the present, secondly use probability-weighted scenarios to address high uncertainty in an explicit way, and thirdly exploit classic analytical techniques to understand the underlying economics of these companies in order to forecast their future performance. Let's examine these stages, one by one.

Stage 1: start from the future

In forecasting the performance of high growth companies the authors stress that analysis should not be constrained by concentrating on current performance. Instead of starting from the present, the usual practice in DCF valuations, one should start by thinking about what the industry and the company could look like when they evolve from today's very high-growth, unstable condition to a sustainable, moderate growth state in the future, and then extrapolate back to current performance. The future growth state should be defined by metrics such as the penetration rate, average revenue per customer and sustainable gross margins. Just as important as the characteristic of the industry and company in this future state is the point when it actually begins. Since Internet-related companies are new, the authors suggest more stable economics probably lie at least 10–15 years in the future.

Combining various assumptions enables forecasts of revenues, profits, capital requirements etc. to be made. The projected free cash flow is then discounted back to give the present value calculations of a company.

Stage 2: weighting for probability

Uncertainty is the hardest part of valuing high-growth technology companies, and the use of probability-weighted scenarios is a simple and straightforward way to deal with it. The use of probability-weighted scenarios requires one to repeat the process of estimating a future set of financials for a full range of scenarios, some more, some less optimistic. The discounted cash flows multiplied by the probability of the scenarios occurring gives us an expected value figure. Sensitivity analysis should then be applied to these scenarios.

Stage 3: customer value analysis

The last difficult aspect of valuing very high-growth companies is relating future scenarios to current performance. How can you tell a soon-to-be successful Internet company from a soon-to-be bankrupt one? Here, classic microeconomic and strategic skills play a critical role because building sound scenarios for business requires knowledge of what actually drives the creation of value. Many of these principles revolve around applying the Porter model, suitably modified, as we discussed in Chapter 5. For many Internet companies, the authors suggest customer value analysis is a useful approach. Five factors are important in applying customer value analysis:

(1) The average revenue per customer per year from purchases by its customers, as well as revenues from advertisements on its site and from retailers that rent space on it to sell their own products.
(2) The total number of customers.
(3) The contribution margin per customer (before the cost of acquiring customers).
(4) The average cost of acquiring a customer.
(5) The customer churn rate (that is, the proportion of customers lost each year).

By using this adapted DCF approach the authors indicate that reasonable valuations can be generated for seemingly unreasonable businesses. Writing before the dot.com crash, the authors correctly stress that uncertainty about the future – a fact of life that cannot be eliminated no matter what type of valuation methodology is applied – should be factored into all valuation methodologies, particularly so for high-risk, technology-related companies.

Conclusion

Early gauges of success, such as how many 'hits' a web page gets, are considered as irrelevant these days. Therefore analysts are trying to set new quantifiable standards, such as lifetime value per user, which are becoming quite popular. Warren Buffet, a very successful value-based investor, recently stated that if he were teaching an 'Investments' class, the final exam question he would set would be 'How do you value an Internet company?'. He supposedly claimed that he would fail anyone who handed him in an answer. The moral of this possibly apocryphal story is that it is very hard to value an Internet company at such an early stage in the industry's life cycle. Successful valuations may identify undervalued and overvalued Internet stocks. However, they cannot address the most important question as to whether the whole industry is overvalued. Recent Nasdaq developments indicate that this question remains unresolved.

10

Derivatives markets, real options and the valuation of Internet and technology stocks

What are derivatives?

Derivatives are contracts that give one party a claim on an underlying asset (derived from the cash value of an underlying asset) at some point in the future, and bind a counterparty to meet a corresponding liability. The contract might describe an amount of currency, a security, a physical commodity, a stream of payments, or a market index. It might bind both parties equally, or offer one party an option to exercise it or not. It might provide for assets or obligations to be swapped. It might be a bespoke derivative combining several elements. Whether derivatives are or are not traded on exchanges; their market price will depend in part on the movement of the price of the underlying asset since the contract was created.

The rapid growth of derivatives trading around the world in recent years has been propelled by the internationalization of capital markets in general, by technological advances in computers and telecommunications, and by the increasingly fierce competition among big banks and securities houses to devise and sell products.

Trading in derivative contracts has a long history. The first recorded accounts of derivative contracts can be traced back to the philosopher Thales of Miletus in ancient Greece, who, during winter, negotiated what were essentially call options on oil presses for the spring olive harvest. De la Vega reported in 1688 that options and futures, or 'time bargains' as they were then known, were trading on the Amsterdam Bourse soon after it was opened. Evidence also suggests that futures contracts for rice were traded in Japan in the seventeenth and eighteenth centuries.

The recent revolution in option pricing theory dates to 1973, with the publications by American financial economists Fischer Black and Myron Scholes of their classic paper on option valuation. Since then, the valuation of options and various other derivative contracts has been one of the primary areas of research among financial economists.

Some terminology

Derivatives are often described as being complex instruments that defy understanding for the mathematically unsophisticated. Despite their intimidating appearance, they are in fact constructed from simple elements, known to the financial markets for centuries.

Take the most basic of derivative transactions, a forward contract. One party agrees to buy, say, $1m in 3 months' time, at a price fixed today in sterling terms. The mathematics of the transaction is well within the capacity of a pocket calculator. If prevailing interest rates are higher for the dollar than for sterling, somebody who wants to buy dollars forward for payment in sterling will be quoted a price lower than the one that is prevailing for transactions that are settled immediately.

Futures contracts differ from forward contracts by virtue of being traded on official exchanges. To make trading easier, their terms will be standard ones set by the appropriate exchange; they will be for a fixed quantity (of bonds, or pork bellies or whatever instrument is being traded) and will run for a fixed period.

Options are a form of forward contracts in which the buyer can decide whether or not to exercise a right to buy (or sell) the underlying asset within an agreed time. The seller of the option then has to work out how to price the probability that the option will or will not be exercised. Only in 1973 did Scholes and Black provide a plausible answer to the option pricing problem by devising a mathematical model with several inputs, the most important of which was the volatility of the price of the underlying asset.

Swaps complete the simple taxonomy. Albeit on a rather larger scale, an interest-rate swap works just as if, for sound financial reasons, Person A with a fixed-rate mortgage, and Person B with a floating-rate mortgage of the same size, agree to assume responsibility for one another's interest payments. Person A will take over the floating-rate payments and person B will take over the fixed-rate payments. In real life, big borrowers may swap interest rate or currency obligations because they disagree over interest rate trends, or because they find it cheaper to borrow money in foreign markets. A Japanese company wanting long-term Japanese yen may find it cheaper to borrow US dollars, then swap them into yen.

Box 10.1 provides a formal definition of the principal derivatives contracts traded: forwards, futures, options and swaps.

What are options?

Options are one of the most powerful derivatives contracts and it is essential to become familiar with the option terminology.

As defined above, an option is a contract between two parties that gives the buyer the right but not the obligation to buy or sell a specific quantity of a commodity or instrument at an agreed price for a specified period. The option buyer pays the seller a premium for the privilege of being able to buy or sell the instrument, at a fixed price, without having the commitment to do so.

To take an example, consider an option to buy gold at US$400 per ounce. Let us say the market price of gold is currently US$395. The option buyer pays the option seller a premium of US$3.50. The option buyer has the right, but not the obligation to buy gold at US$400. It will be profitable for the option buyer to exercise this right if the price of gold rises above US$403.50. However, if the price of gold falls in the market then the option buyer has no commitment to buy gold at US$400, and the option buyer can then allow the option to expire unexercised, and purchase gold at the cheaper market price.

Some terminology

A summary of the principal terminology of the options market is offered below:

Box 10.1 Derivatives defined

Forward contract: A contract to buy or sell a specified amount of a designated commodity, currency, security, or financial instrument at a known date in the future and at a price set at the time the contract is made. Forward contracts are negotiated between the contracting parties and are not traded on organized exchanges.

Futures contract: A contract to buy or sell a specified amount of a designated commodity, currency, security, or financial instrument at a known date in the future and at a price set at the time the contract is made. Futures contracts are traded on organized exchanges and are thus standardized. The contracts are marked to market daily, with profits and losses settled in cash at the end of the trading day.

Option contract: A contract that gives its owner the right, but not the obligation, to buy or sell a specified asset at a stipulated price, called the strike or exercise price. Contracts that give owners the right to buy are referred to as call options and contracts that give the owner the right to sell are called put options. Options include both standardized products that trade on organized exchanges and customized contracts between private parties.

Swap contract: A private contract between two parties to exchange cash flows in the future according to some prearranged formula. The most common type of swap is the 'plain vanilla' interest rate swap, in which the first party agrees to pay the second party cash flows equal to interest at a predetermined fixed rate on a notional principal. The second party agrees to pay the first party cash flows equal to interest at a floating rate on the same notional principal. Both payment streams are denominated in the same currency. Another common type of swap is the currency swap. This contract calls for the counterparties to exchange specific amounts of two different currencies at the outset, which are repaid over time according to a prearranged formula that reflects amortization and interest payments.

- The option buyer becomes the **holder**. The option seller is called the **writer**.
- A **call** option gives the owner the right to buy a specified quantity of a commodity at an agreed price over a given period.
- A **put** option gives the owner the right to sell a specified quantity of a commodity at an agreed price over a given period.
- The **premium** is the price paid for the option.
- The **strike price** or **exercise price** is the rate at which the option may be exercised; in other words, it is the price which has been agreed under the option contract.
- The **expiry date** is the final date on which the option can be exercised.
- A **European-style** option can be exercised only on the expiry date, whereas an **American-style** option can be exercised at any date prior to and including the expiry date. (Note that these terms have no geographical significance.)

Exchange-traded versus over-the-counter options

Options may be traded on exchanges, i.e. in a physical location, or on the over-the-counter (OTC) market, in which dealing takes place between two counterparties, usually over the telephone.

Exchange-traded options have the following characteristics:

- Fixed expiry dates, generally at 3-monthly intervals for the third Wednesday in March, June, September and December.
- Maturities generally up to two years.
- Strike/exercise prices at fixed intervals.
- Fixed contract sizes.
- Standardization of contracts. This means that markets tend to be liquid. In other words, bid-offer spreads (the difference between the buying and selling price) tend to be narrow, and large orders can usually be transacted fairly easily.
- Given that these options are traded on regulated exchanges, trading is closely monitored. The clearinghouse of the exchange acts as the counterparty to every trade, thus the credit risk, i.e. the risk of default on a trade, is standardized and limited.
- Prices are publicly quoted, i.e. trading takes place by open outcry between traders on the floor of the exchange. Prices

are reported by information vendors such as Bloomberg and Reuters, and prices and volumes are reported in the financial press.

Over-the-counter options have the following characteristics:

- Strike rates, contract sizes and maturity are all subject to negotiation.
- They can be longer term than exchange-traded options – some banks will write them for up to 10 years.
- The holder has a direct credit risk on the writer. The writer has no credit risk on the holder provided the premium is paid up front.
- The price at which the option is dealt is known only to the counterparties.

Who participates in the derivatives markets?

In order to understand how to apply these arbitrage principles, and ultimately to understand how option prices are determined, we need to examine the actions of the different participants in the markets. In this case we choose the gold market, but any other market could easily have been selected.

The expected future price of gold is important to anyone who uses the gold *spot* markets, that is, the market for immediate delivery or receipt of gold, and to the gold *futures* and *forward* markets, that is, the markets for delivery at some future time period. These users would include:

(A) Investors
(B) Long speculators
(C) Short speculators
(D) Long hedgers
(E) Short hedgers
(F) Arbitrageurs

It is useful to describe each of these market participants, discuss what motivates their market behaviour, and look at how their actions affect the spot and the future or forward price of an underlying commodity or security.

Let's look first at our investor, A. He thinks that gold is a good investment – a 'gold bug' in the terminology of the market. He wants to buy gold and hold it long term. When A buys gold, he forces up the spot price of gold. When the spot price of gold rises, the future or forward price tends to rise as well.

B, our long gold speculator, is not a long-term investor, like A, but she thinks gold is going up and would like to take a position that is going to be profitable if she is right. B can accomplish this in a couple of different ways. First, B can buy spot gold. If gold goes up in a short period of time, she can sell that gold at a higher price and make a profit.

Alternatively, B can buy a future or forward contract on gold. B's purchase of forward gold tends to make the future or forward price go up; and, just as demand for spot gold tends to raise future or forward prices, demand for forward gold tends to raise spot prices. As we will see later, when B buys a future or a forward, she does not have to pay out much money. When A buys spot gold, say 100 ounces at $400, he has to pay $40 000, immediately.

Our short speculator, C, thinks that the gold price is going down. She is very pessimistic about the outlook for gold. Like A and B, C has two choices. She can sell gold spot, or she can sell a future or forward contract on gold.

If C happens to own gold, she can simply sell it. Alternatively, if she is in a position to borrow gold relatively easily and sell it short, i.e. to sell gold she does not physically own, she might do that. But if she does not have direct access to gold, she probably will find it easier to sell a future or forward contract on gold.

Short speculators like C affect the price of spot gold. The effect on the spot price is clear if the short speculators own gold and want to shift their investment to something else. If they are not trading in the spot market, the direct effect of their sale will be on the forward or future price, making it trade at a lower price.

D is someone who needs gold a year from now. D might be a dentist or a jeweller – someone who has all the gold he needs right now but who wants to be assured of the gold price a year from now. There is no advantage in buying the gold now and spending a large sum of money to store and insure it. D probably will want to buy a future or forward contract on gold, which will tend to push up the future or forward price.

E is a gold producer, who is very busy digging gold out of the ground. He does not have much gold right now, but in a year he expects to have a large store of gold to sell. He wants to make sure his hard work pays off and that he can sell his gold at a good price. He cannot sell spot gold because he does not have it. However, he can sell a future or forward contract on gold now. E is likely to affect the future or forward price, and since he is selling, his actions will tend to make the future or forward price trade lower.

Our last market participant, the gold arbitrageur, is the most important in some ways. F has no opinion about gold's value and no opinion on the likely direction of gold prices, but she is very aware of gold price relationships. Even if F did have a personal opinion about the value of gold or the direction of gold prices, her opinion would not affect her financial transactions. As an arbitrageur, it is F's actions in the market place that tie the actions of all other market participants together. F makes sure the spot price of gold and the future or forward price of gold are in their proper relationship. If the prices are not appropriate relative to one another, she will try to profit by buying in the cheap market and selling in the expensive market. The actions of arbitrageurs are extremely important in seeing that spot prices and future prices are kept in line.

The role of the arbitrageur and the pricing of futures markets

One of the reasons for emphasizing the role of the arbitrageur in a discussion of options is that arbitrageurs are an important factor in determining the spot/forward price relationship, and the first step in finding the value of an option is finding the value of future or forward prices.

If gold is trading at $400 today, at what price will someone agree today to buy or sell it one year in the future? There is a rational, deterministic relationship between the spot price and the forward price of gold that does not rely on an opinion survey or on anyone's gold price forecast, it simply applies arbitrage ideas discussed above.

Suppose that the carry cost of gold – that is, the cost of borrowing money, the cost of buying the gold and financing it for a year – is 10%. Assume the spot gold price is $400. Holding the gold position will cost us 10% for a year, because if we had to borrow the $400 to buy the gold we would have had to pay interest at 10%. To own gold a year from now, it will cost us more than \$400 now. In fact as shown by equation 10.1, it will cost us \$440.

$$\text{Spot price of gold} + \text{Cost of carry} = \text{Future or forward price of gold}$$

$$\$400 + (10\% \times \$400) = \$440 \qquad \text{Eq 10.1}$$

If the present value of gold is \$400, the future value of gold in one year should be \$440. By future value, we do not necessarily mean the price gold will sell for in the future; it refers to the value that the future gold price has today because of the cost of carrying gold for a year.

Present value, spot value and the cash market refer to the same thing. We have been talking about spot gold, but we can also talk about the spot value, current market value or cash value of a stock, a bond, or some other *underlying* commodity or financial instrument, to broaden the futures pricing principles.

Future value is the value of the future or the forward. There is a difference between futures and forwards, but the difference is in how they are traded, not in how they are valued. At this point we will treat futures and forwards as though they were interchangeable. Later, we will discuss the practical differences between futures markets and forward markets.

Factors influencing the price of futures

Traders in any commodity, say a wheat buyer (bread maker), or a wheat seller (wheat farmer), fearing wheat price volatility, are anxious to use a cash market hedge to hedge wheat price risk. A futures contract is a derivative of this spot market hedge. A seller of wheat will not wish the price to fall in the future when a sale is anticipated. A buyer of wheat will not wish the price to rise when a purchase is anticipated. If an agreement is made to deliver a quantity and grade of wheat in 3 months' time, how might the seller of the wheat arrange affairs so that there is no price risk? If they remain in an open position, i.e. not owning the wheat, there is no knowing what the price of wheat will be in 3 months' time. Therefore a cash market hedge can be constructed as follows. Wheat will already be held or can be purchased at today's spot price. Take today's spot price, which is known, and add to this the **cost of carrying** the wheat for three months. These carry costs will be:

- storage
- insurance
- transport costs involved in making delivery to a named place and
- financing costs of the operation over the 3 months, i.e. interest foregone, or interest paid on funds used to purchase the commodity.

This gives us the following relationship between the spot and futures price:

> spot wheat price + the cost of carry = agreed price of wheat in 3 months' time (i.e. the futures price)

Such a strategy will fix the price of wheat for both parties in 3 months' time. Regardless of what happens to the spot price during the 3 months, the agreed price will be received or paid.

Agreeing the futures price

A price in the futures agreement or contract has to be agreed. The question arises as to what must/should this price be?

The price, as indicated above, 'should' be today's spot price adjusted by the cost of carry for 3 months. If you agree this price with your counterparty this will become the entry price, as it is known. If the entry price is not 'correct', then an arbitrage can be made, as illustrated below, using the cash market hedge described above and the simultaneous mispriced futures agreement.

If the entry price of the futures contract (remember this will give the seller of wheat the right to sell the wheat at this agreed entry price in 3 months' time) is greater than spot plus cost of carry then an arbitrage profit is possible. In this example, the futures price is said to be 'rich' to the cash price, so on the principle that you sell that which is overpriced (the future) and buy that which is underpriced (commodity at spot), it follows that:

If future (entry) price		>	spot plus cost of carry
(e.g.	$10	>	$8 + $1.50)

then arbitrageurs in the spot market will take a long position, i.e. purchase wheat at the spot rate , and carry this for 3 months.

Their total outlay	=	spot price + the cost of carry
\$9.50	=	\$8 + \$1.50

In the futures market they will take a short position, i.e. sell a futures contract. In 3 months' time they will sell the goods held over the 3 months at a total outlay of original spot (\$8) plus cost of carry (\$1.50), for a sum greater than this (\$10), as enabled by the futures contract held, because:

future contract price	>	spot price plus cost of carry
\$10	>	\$8 + \$1.50

This arbitrage process is called a **cash and carry transaction**. In this example a risk-free profit is being made. The arbitrage opportunity will be eroded and disappear as such transactions are made. Demand at the spot price will raise the spot price so that this, when added to the cost of carry, equals the futures price. At the same time there will be a greater demand for short positions in futures and this will drive down the futures price. In both markets the tendency will be to equalize prices. If the futures price equals the spot price plus full carrying cost, then the futures price will be a **full carry price**.

If the futures price is 'cheap', then arbitrageurs buy what is cheap (the futures market) and sell that which is expensive (the spot price). This arbitrage again 'should' result in a **full carry futures price**.

At the end of either of these transactions the futures entry price should equal the spot price plus the cost of carry.

The principle emerges therefore that spot and future prices differ due to the principle of cost of carry. However, at maturity, spot and futures will be identical. If they are not identical it will be possible, a fraction of time before maturity, to take a futures position and an opposite cash market position, arbitrage and profit from the difference. A simple example illustrates this.

Say the futures price is \$10 and the spot price is \$8. Selling a futures contract (right to deliver at \$10) and buying in the spot market at \$8 obviously gives a profit of \$2 per unit quantity. If prices are identical at maturity, but differ at the beginning of the period, it follows that even if the spot price were to remain constant (highly unlikely) the futures price would gradually have to change as the contract approached maturity. To illustrate this we must first define what in the derivatives market is known as 'basis'.

What is basis?

Basis is the difference between the forward or future value and the spot value. For many products, the basis is defined as the cost minus the benefit of holding the underlying security. This gives the following relationship:

basis = cost of transaction − benefit from the transaction

In our gold futures pricing example, we considered only the financial cost of carrying gold at 10%, because there are no economic or financial benefits from owning gold. (It might give you a feeling of comfort to own gold, but we are not counting that.) In some other products, there are benefits to holding financial instruments as well as costs.

If we now consider the futures markets for stocks we must again consider the cost of carry, that is, the cost of the money to buy the stock, but we must now also consider the possible benefit of holding stocks, i.e. the possibility of receiving a cash dividend. Not all stocks pay dividends, but if they do, the dividend is a benefit.

In currency markets, the cost of carrying a foreign currency is the investor's domestic interest rate. If US dollar-based investors wants euros, they have to give up dollars to buy the euros. When they give up the dollars, they either take them out of an interest-bearing account, which means they lose the interest income, or they borrow the dollars from a bank and pay domestic interest. Either way they have a carry cost in the domestic currency, and that is the dollar cost. On the other hand, when they get the euros, they can invest them in a euro account and earn interest at euro rates. The interest on the euro account is a benefit of owning the euros.

Let us now turn to the bond market. Often, traders or portfolio managers have to borrow money to carry bonds. They borrow money in the *Repo (repurchase agreement) market*, posting the bonds as *collateral* for the loan. The *Repo rate* is the term used for the interest rate charged on such a loan. On the other hand, while they own the bond they are entitled to any coupon interest that accumulates. For a bond, the Repo rate is the carry cost of the bond, and the coupon interest is the benefit.

The basis in each of these examples consists of the cost of holding the instrument or commodity minus the benefit of owning the underlying instrument or commodity from the spot date to the forward date.

The basis for different instruments is as follows:

- For gold, the basis is simply the interest cost of carrying the gold.
- For a stock, the basis is the interest cost of carrying the stock minus the dividends earned from the stock.
- For a currency, the basis is the cost of the interest on the domestic currency minus the benefit of the interest earned on the foreign currency.
- For a bond, the basis is the cost of borrowing at the Repo rate minus the benefit of the coupon payment.

What are forward market contracts?

A forward contract is an agreement between two parties made independently of any organized exchange market. Nobody other than the parties to the forward agreement needs to be involved – there are no formal rules outside the agreement between the two parties. The forward contract is a stand-alone, customized contract. It can be based on any amount of any good. It can be written for settlement at any time and at any price. It can be, as an extreme example, for delivery of 37 000 gallons of vodka in 52 days at $2 a gallon. The terms can be virtually anything as long as both parties agree to them.

One party agrees to sell the vodka at the contract price, and the other party agrees to buy it at that price. Since the contract can be for any amount of any good, at any time and at any price, and since each party depends on the other party to meet contractual obligations, the forward contract *cannot* be traded freely with other potential counterparties. Forwards are not *fungible*; that is, one forward contract is not interchangeable with another contract that has similar terms, and the present value of the forward contract is not easily converted to a cash market value.

Each party has to be able to trust that the other party will uphold their side of the agreement, because the two counterparties are exposed to each other's ability and willingness to perform on the contract. There is *credit risk* associated with a forward contract, which must be controlled. This is based in the fact that you cannot be sure that the counterparty will deliver as agreed. Often, the credit risk is controlled by banks, which act as *intermediaries*, assuring the performance of their clients. A very important point to remember about forward contracts is that no money is exchanged by the parties until the actual exchange of goods or financial instruments at *settlement*. There is normally no interim cash flow.

What are futures contracts?

A futures contract, in contrast to a forward, is an agreement between two parties made through their agents on an organized futures exchange. The parties who trade on the exchange can represent themselves or they can represent customers. There are specific rules and regulations that set the terms of the contract and the procedures for trading. The contract is for a specific amount of a specific good to be delivered at a specific time determined by the exchange, and the price discovery usually, but not always, is determined by *open outcry* in a *trading pit*, that is, by people yelling prices at each other in a room.

This is the way prices are discovered and goods are exchanged in many efficient markets. Some markets now use computer systems, known as screen-based trading, which expose bids and offers to a large number of potential traders at many locations, but the principle of bringing bids and offers together is the same.

Several features of futures markets are worth noting. First, a futures contract is always for a specific amount of a specific good. We cannot set out our own contract amount, say 37 000 gallons of vodka. If the vodka contract set by the exchange is 20 000 gallons, we can trade any number of contracts, but each contract must be for 20 000 gallons. And we must specify the type of vodka – it must be Finlandia vodka or Stolichnaya vodka, or, more likely, simply 80-proof vodka. Since all terms are determined in the futures contract, the contracts can be retraded freely with other counterparties. We can buy the contract from one person, sell the contract to another, and wash our hands of the commitment. We can eliminate our obligation, because, with futures, the organized futures exchange handling the transaction takes the other side of the contract.

The futures exchange is the ultimate counterparty for all futures trades, so the only credit risk is the credit-worthiness of the exchange, which is normally very low. Once a trade is completed, the transaction is passed to some type of exchange clearing corporation. The futures contract buyers and sellers have agreement with the clearing corporation. Ultimately, if a trader takes delivery rather than offsets the contract, i.e. reverses it, the exchange will select someone who is what is technically known as 'short' the contract to make delivery. The party making delivery does not have to be the party who sold the contract originally. After the trade settles, the original parties to the trade lose any direct tie to each other based on the trade. No credit intermediary is necessary, but margin must be deposited to ensure each party meets its obligations.

The exchange clearing corporation has to stand behind the credit-worthiness of its members, so it asks everyone for a deposit. Customers make a deposit with their broker, known as margin, and the broker passes the deposit to the clearing corporation. Futures margin is a good faith deposit, demonstrating ability and willingness to meet contractual obligations. Table 10.1 illustrates the advantages and disadvantages of futures and forward contracts.

Table 10.1 The advantages and disadvantages of futures and forwards

	Futures markets	*Forward markets*
Default risk	+ Low	– Greater than for futures
Transaction costs/ commissions	+ Low	– Higher than for futures
Standardization	+ Contracts are liquid; allows for secondary market	– No secondary market
	– Imperfect hedge	+ Tailor-made hedge
Interest rate risk	– Daily cash flows from marking to market must be deposited at unknown interest rates	Not applicable
Contract sizes and underlying currencies	– Limited number of contracts and underlying securities makes hedging with futures less effective	+ Tailor-made
Maturities	– Short maturity only	+ Slightly longer to much longer maturities than for futures

How are options priced?

The method used to price options depends on the type of option being priced. As can be seen from Table 10.2, the two pricing models are the binomial model used for American options, which was designed by Cox, Ross and Rubinstein, and the Black–Scholes option pricing model, used for pricing European options. As we show below, the binomial model can also be used to price European options.

Table 10.2 European versus American options

Type of option	*Exercise date*	*Pricing model*
American option	Any time up to maturity	Cox, Rubinstein and Ross
European option	At maturity date	Black–Scholes

The key to valuing options is to design a risk-free portfolio in which the value does not change when there is a change in the underlying asset. This risk-free element is achieved by designing a portfolio which replicates, i.e. one in which the payoffs of the portfolio exactly match the payoffs of the option. A good example of this is achieved by applying the binomial model.

The binomial model

One-period binomial model – the role of the replicating portfolio

To illustrate the binomial model it is necessary to look at what the effect of the prices of the underlying stock rising or falling in price should have on the option price. Our first example is a European call option on a stock, and assuming that the stock is currently valued at $100. In this example an option to purchase this stock expires in one year and the strike or exercise price is $100. The annual risk-free interest rate is 5%, so that borrowing $1 today will mean having to pay back $1.05 one year from now. For simplicity, the assumption here is that there are only two possible outcomes when the option expires – the stock price can be either $120 (an up state) or $80 (a down state). Note that the value of the call option will be $20 if the up state occurs and

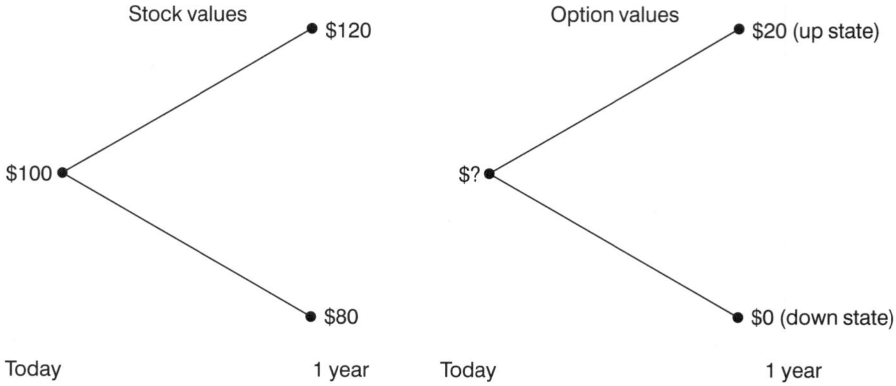

Figure 10.1 Stock and option values in the one-period model

\$0 if the down state occurs as shown in Figure 10.1. In other words, if the price falls to \$80 the option becomes valueless.

Since there are only two possible states in the future it is possible to replicate the value of the option in each of these states by forming a portfolio of the stock and risk-free asset. If Δ, an unknown number of shares of the stock are purchased and M, an unknown number of dollars, are borrowed at the risk-free rate, the stock portion of the portfolio is worth $120 \times \Delta$ in the up state and $80 \times \Delta$ in the down state while $1.05 \times M$ will have to be paid back in either of the two states. Thus, to match, or replicate, the value of the portfolio to the value of the option in the two possible states, it must be the case that

$$\$120 \times \Delta - 1.05 \times M = \$20 \text{ (up state)} \qquad \text{Eq 10.2}$$

and

$$\$80 \times \Delta - 1.05 \times M = \$0 \text{ (down state)} \qquad \text{Eq 10.3}$$

By rearranging the formulae from Equations 10.2 and 10.3 gives us a value for Δ:

$$\$40\Delta = 20$$

$$\therefore \quad \Delta = 0.5$$

Inserting $\Delta = 0.5$ in Equation 10.2 gives us

$$\$120 \times 0.5 - 1.05M = \$20$$

$$\$40 = 1.05M$$

So $M = \$38.10$.

The resulting system of two equations with two unknowns Δ and M, solved above, gives us $\Delta = 0.5$, and M is approximately \$38.10. This means that one would need to buy 0.5 shares of the stock and borrow \$38.10 at the risk-free rate in order for the value of the portfolio to be \$20 and \$0 in the up state and down state, respectively. Equivalently, selling 0.5 shares of the stock and lending \$38.10 at the risk-free rate would mean payoffs from that portfolio of –\$20 and \$0 in the up and down state, respectively, which would completely offset the payoffs from the option in those states. This is a very powerful finding as it enables us to replicate the risk-free portfolio and price the options accordingly. It is also worth noting that the current value of the option must equal the current value of the portfolio, which is $100 \times \Delta - M = 100 \times 0.5 - M = \$11.90$. In other words, a call option on the stock is equivalent to a long position in the stock financed by borrowing at the risk-free rate.

The variable Δ is called the *delta of the option*. In the previous example, if C_u and C_d denote the values of the call option in the up and down states, with S_u and S_d denoting the price of the stock in the up and down states, respectively, then it can be verified that $\Delta = (C_u - C_d)/(S_u - S_d)$. The delta of an option reveals how the value of the option is going to change with a change in the stock price. For example, knowing $\Delta = C_d$, and the difference between the stock prices in the up and down state makes it possible to know how much the option is going to be worth in the up state, that is to say that C_u is also known.

The two-period binomial model

A model in which a year from now there are only two possible states of the world is certainly not realistic, but construction of a multi-period model can alleviate this problem. The one-period model assumes a replicating portfolio for a call option on a stock currently valued at \$100 with a strike price of \$100 and which expires in a year. However, let us now assume the year is divided into two 6-month periods and the value of the stock can either increase or decrease by 10% in each period. Assume the semi-annual risk-free interest rate is 2.47%, which is equivalent to an annual compounded rate of 5%. The states of the world for the stock values are given in Figures 10.2. and 10.3. Given this structure, how does one build a portfolio of the stock and the risk-free asset to replicate the option? The calculation is similar to the one above except that it is done recursively, starting one period before the option expires and working backward to find the current position.

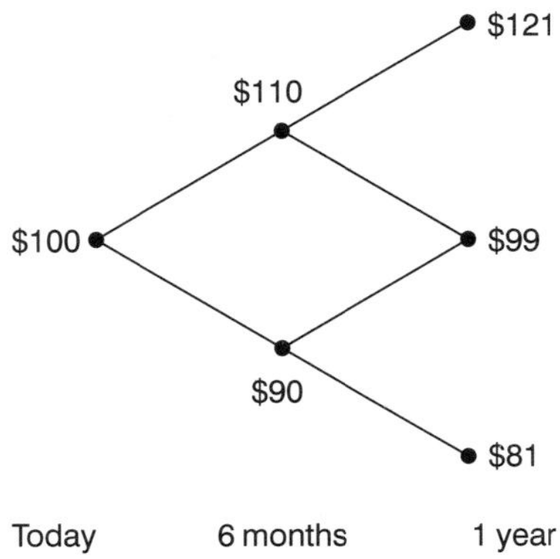

Figure 10.2 Stock values in the two-period model

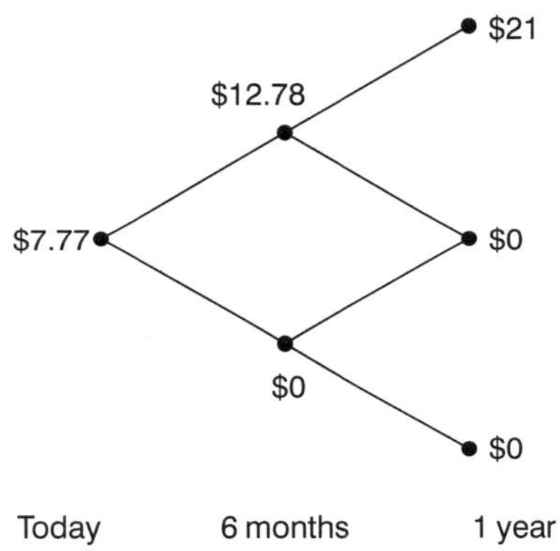

Figure 10.3 Option values in the two-period model

In the case in which the value of the stock over the first 6 months increases by 10% to $110 (that is, the up state 6 months from now), the value of the option in the up state is found by forming a replicating portfolio containing Δ_u shares of the stock financed by borrowing M_u dollars at the risk-free rate. Over the next 6 months, the value of the stock can either increase another 10% to $121 or decline 10% to $99, so that the option at expiration will be worth either $21 or $0. Since the replicating portfolio has to match the values of the option, regardless of whether the stock price is $121 or $99, the Equations 10.4 and 10.5 must be satisfied:

$$\$121 \times \Delta_u - 1.0247 \times M_u = \$21 \qquad \text{Eq 10.4}$$

and

$$\$99 \times \Delta_u - 1.0247 \times M_u = \$0 \qquad \text{Eq 10.5}$$

Solving these equations results in $\Delta_u = 0.9545$ and $M_u = 92.22$. Thus the value of the replicating portfolio is $110 \times \Delta_u - M_u =$ \$12.78. If instead, 6 months from now the stock declines 10% in value, to \$90 (the down state), the stock price at the expiration of the option will either be \$99 or \$81, which is always less than the exercise price. Thus the option is worthless a year from now if the down state is realized 6 months from now, and consequently the value of the option in the down state is zero. Given the two possible values of the option 6 months from now, it is now possible to derive the number of shares of the stock that one needs to buy and the amount necessary to borrow to replicate the option payoffs in the up and down state 6 months from now. Since the option is worth \$12.78 and \$0 in the up and down states, respectively, it follows that

$$\$110 \times \Delta - 1.0247 \times M = \$12.78 \qquad \text{Eq 10.6}$$

and

$$\$90 \times \Delta - 1.0247 \times M = \$0 \qquad \text{Eq 10.7}$$

Solving Equations 10.6 and 10.7 results in $\Delta = 0.6389$ and $M =$ \$56.11. Thus the value of the option price today is $100 \times \Delta - M$ = \$7.77. The values of the stock option are shown graphically in Figures 10.2 and 10.3.

A feature of this replicating portfolio is that it is always self-financing. Once it is set up, no further external cash inflows or outflows are required in the future. For example, if the replicating portfolio is set up by borrowing \$56.11 and buying 0.6389 shares of the stock and in 6 months the up state is realized, the initial portfolio is liquidated. The sale of the 0.6389 shares of stock at \$110 per share nets \$70.28. Repaying the loan with interest, which amounts to \$57.50, leaves \$12.78. The new replicating portfolio requires borrowing \$92.22. Combining this amount with the proceeds of \$12.78 gives \$105, which is exactly enough to buy the required 0.9545 Δ_u shares of stock at \$110 per share. Replicating portfolios always have this property: liquidating the current portfolio nets exactly enough money to form the next portfolio. Thus the portfolio can be set up today, rebalanced at the end of each period with no infusions of external cash, and at expiration should match the payoff of the option, no matter which states of the world occur.

In the replicating portfolio presented above, the option expires either one or two periods from now, but the same principle applies for any number of periods. Given, in this example, that there are only two possible states over each period, a self-financing replicating portfolio can be formed at each date and state by trading in the stock and a risk-free asset. As the number of periods increases, the individual periods get shorter so that more and more possible states of the world exist at expiration. In the limit, continuums of possible states and periods exist so that the portfolio will have to be continuously rebalanced. The Black–Scholes–Merton model, discussed below, is the limiting case of these models with a limited number of periods.

Determining the value of call options

A call option is the right to purchase the underlying asset for a specified exercise price until the expiration date. As discussed earlier, American options can be exercised at any time. European options can be exercised only at the expiration date. In the discussion below, which draws on Livingston (1996), the following notation is used:

C = market value of the call option
P = market value of the underlying asset
E = exercise price (strike price)

If the price of the underlying asset (P) is less than the exercise price ($P < E$), the call is described as being *out-of-the-money*. If P equals E, the call option is described as being *at-the-money*. If the price of the underlying asset exceeds the exercise price ($P > E$), the call option is described as being *in-the-money*.

What is the value of the call option at expiration?

At expiration, the value of the call option must be zero if the market value of the underlying asset is less than or equal to the exercise price. No rational investor would exercise a call option if the underlying asset sells for the exercise price or less since buying the underlying security in the open market would be cheaper than exercising the call option. For example, in Table 10.3, if a bond sells for $90 in the open market, a call option with an exercise price of $100 is valueless at expiration. Instead of exercising the call and paying the $100 exercise price, it is preferable to buy the bond directly for $90.

In Table 10.3, if the price of the underlying asset exceeds the exercise price, the call option is worth the price of the

Table 10.3 Value of call option at expiration (E = $100)[tacpr]

$P<E$	$P=E$	$P>E$
e.g., $P = 90$	$P = 100$	$P = 110$
$C = 0$	$C = 0$	$C = P - E$
		e.g. $C = 10$
out-of-the-money	at-the-money	in-the-money

Source: Livingston, 1996

underlying asset minus the exercise price, that is, $P - E$. The amount $P - E$ is called the *intrinsic value* of a call option. If the call option sells for less than P – E, an arbitrageur would buy the call, exercise it and make an arbitrage profit. At expiration, buying the call option for $\$C$ and exercising it is equivalent to buying the underlying security directly for $\$P$. Thus, $P = C + E$, or $C = P - E$.

At expiration, the value of a call option is:

$$C = 0 \qquad \text{if } P \leq E \quad \text{at- or out-of-the-money} \qquad \text{Eq 10.8}$$

$$C = P - E \qquad \text{if } P > E \quad \text{in-the-money} \qquad \text{Eq 10.9}$$

What is the profit profile for a call option?

The possible value of a call option can be seen from a profit profile. As a reference point, consider someone who buys a bond (for which options trade) currently selling at its exercise price of $100 and holds this bond for 3 months until the option expires. The possible profits and losses, overlooking coupon interest, are shown as a solid line in Figure 10.4.

Consider the purchase of a call option for $4 with an exercise price of $100. The investor holds this call option for 3 months until expiration. The profit profile is shown as the dotted line in Figure 10.4. The first step in drawing the profit profile is to compute the profit and loss on the position for several levels of the price of the underlying asset at expiration. These prices should include the option's exercise price and several points on either side of the exercise price. The procedure is illustrated in Table 10.4.

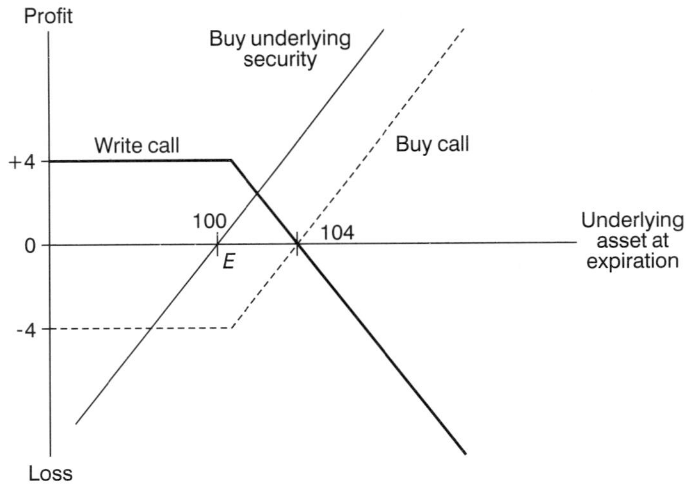

Figure 10.4 Profit profiles for a call option. *Source:* Livingston, 1996

The call buyer suffers a loss of $4, the entire purchase, if the bond price is below the exercise price of $100. This $4 is the maximum loss for the buyer of a call option. If the bond price at expiration is above the exercise price, the profit equals $P - E =$ $4. A net loss is incurred if the bond price is below $104 and a net profit if the bond price is above $104. The profit profile indicates that call buyers are anticipating a rising price for the underlying asset.

If the bond price is at or below the exercise price, the profit profile shows the call buyer's maximum loss to be the purchase price, *C*, since this call purchase price is a small percentage of

Table 10.4 Profits or losses for a call buyer ($E =$ $100)

	Price of underlying security at expiration			
	98	*100*	*102*	*104*
Buy call (C)	–4	–4	–4	–4
Exercise call at expiration			–100	–100
Sell underlying bond acquired from exercising option			+102	+104
Net profit = $+C - E + P$	–4	–4	–2	0

Source: Livingston, 1996

the bond price, the loss from buying a call is small in absolute dollars compared to the loss from purchasing the bond outright. On the other hand, the call buyer gains substantially if the bond does well.

For everyone who buys a call option, someone sells or writes the call. The call writer agrees to sell the underlying asset at the exercise price if the option is exercised by the call buyer. Call options are a zero-sum game, meaning that the call buyer's gains are the call writer's losses and vice versa. In effect, the buyer and the writer are betting against each other.

What are the determinants of the value of a call option?

The value of a call option is determined by the following six factors.

(1) **The price of the underlying asset**: The value of a call option is positively related to the price of the underlying security. The higher the value of the underlying asset, the greater the value of the call option. This is not surprising, as a call buyer has bought the right to buy at a fixed price. The more the underlying asset rises the more the call option is worth, and vice versa.
(2) **The exercise price:** The value of a call option is inversely related to the exercise price. The lower the exercise price, the higher the value of the call option. Other things being equal, a lower exercise price means that the call option is more in-the-money. Again this is not surprising as this is what the call option owner has to pay when he exercises the option.
(3) **The time until expiration**: The value of a call option is a positive function of time until expiration. The longer the time until expiration, the greater the value of the call option. The reason for this is that the call owner has more time to allow his option to get in-the-money.

The first three determinants of call option value are illustrated by Table 10.5. Look across any row of the table; the value of the call option decreases as the exercise price increases. Look down any column; as time to maturity increases and the value of the call option also increases.

(4) **The price volatility of the underlying asset:** The greater the volatility of the underlying asset, the higher the value of

Table 10.5 The impact of maturity and exercise price on call option prices

Maturity (mth)	*Exercise price*[a]		
	$90	*$100*	*$110*
3	$30	$16	$4
6	$34	$19	$6.50
9	$37	$21.50	$8.50

[a]Price of underlying asset = $110.
Source: Livingston, 1996

the call option, since payoffs to a call buyer are asymmetric. If the underlying asset does poorly, the call buyer loses everything. If the underlying asset does well, the call buyer does very well. Greater dispersion in the possible value of the underlying asset implies bigger call option payoffs on the upside but the same payoff (loss of everything) on the downside, making a call option more valuable. Obviously asset prices can fall as well as rise but if they fall the option holder has the option not to exercise his option.

The impact of volatility upon call option value is illustrated by the two-security example in Table 10.6. Assume two securities with the following possible prices at expiration. Each of these securities has the same mean of 100, but security 2 has greater

Table 10.6 Two-security example of volatility

Security 1			
Prices at expiration	90	100	110
Probability	1/3	1/3	1/3
Mean price = (90)(1/3) + (100)(1/3) + (110)(1/3) = 100			
Security 2			
Prices at expiration	80	100	120
Probability	1/3	1/3	1/3
Mean price = (80)(1/3) + (100)(1/3) + (120)(1/3) = 100			

Source: Livingston, 1996

Table 10.7 Value of a call option

Call option on Security 1			
Prices of underlying	90	100	110
Value of call option	0	0	10
Probability	1/3	1/3	1/3
Mean value = (0)(1/3) + (0)(1/3) + (10)(1/3) = 3.33			
Call option on Security 2			
Prices of underlying	80	100	120
Value of call option	0	0	20
Probability	1/3	1/3	1/3
Mean value = (0)(1/3) + (0)(1/3) + (20)(1/3) = 6.67			

Source: Livingston, 1996

dispersion of outcomes. An option on security 2 is more valuable because of this dispersion. To see the point, consider the case of call options with exercise prices of $100 on each security. The payoffs on these options are shown in Table 10.7.

The call option on security 2 in Table 10.6 clearly has a greater value. If the options are out-of-the-money or at-the-money, both options expire worthless. If the options are in-the-money, the second option has a higher payoff and, therefore, must be worth more.

(5) **The risk-free interest rate**: The higher the interest rate, the greater the value of a call option. This is based on the fact that if a call buyer has bought the option rather than the underlying asset then the funds saved can then be invested at the now higher interest rate.
(6) **Dividends or interest on the underlying asset**: The higher the cash payments on the underlying asset the lower the value of a call option. The total return on an asset is the cash payment (dividends or coupon interest) plus price appreciation. For a given total rate of return, higher cash payments on an asset imply lower returns from price increases. Since the call buyer gains only if the price of the underlying asset increases, higher cash payments on the underlying asset tend to reduce the capital gains and the value of the call option.

Table 10.8 Factors affecting option prices

Variable increases	*Call price*	*Put price*
Share price	INCREASES	DECREASES
Time to expiry	INCREASES	INCREASES
Share volatility	INCREASES	INCREASES
Risk-free rate	INCREASES	DECREASES
Exercise price	DECREASES	INCREASES
Dividend	DECREASES	INCREASES

From put-call parity, the price of a put can be shown to depend upon the call price, the price of the underlying asset and the present value of the exercise price. It follows that the preceding six determinants of call prices also affect put prices. There are two major differences for puts. First, as the price of the underlying security increases, the value of the put goes down. That is, the value of a put is inversely related to the price of the underlying security. Second, a higher exercise price increases the value of the put. The value of a put option is directly related to the exercise price.

Table 10.8 provides an illustration of how both call options and put options respond to the factors influencing their price.

The Black–Scholes model

The binomial model, discussed earlier, assumes a discrete-time stationary binomial stochastic process for security price movements. In the limit, as the discrete-time period becomes infinitely small, this stochastic process becomes a *diffusion process* (also called a *continuous-time random walk*, an *Ito process*, or *geometric Brownian motion*). This was the process assumed by Black and Scholes (1973) in their derivation of the option pricing formula. As with the binomial model, Black and Scholes begin by constructing a riskless hedge portfolio, long in the underlying security and short in call options. This portfolio generates the riskless rate of return, but the internal dynamics of the portfolio are driven by the diffusion process for the security price. The structure of the hedge portfolio can be put into a form that is identical to the heat equation in physics.

Once this was recognized, the solution to the equation was easily derived.

In the case of a European call option with no cash payments on the underlying asset and with a certain, continuously compounded interest rate, Black and Scholes demonstrated that the value of a call option is an explicit function of the first five factors mentioned above. The Black–Scholes model is shown in Equation 10.10:

$$C = PN(d_1) - Ee^{-rt} N(d_2) \qquad \text{Eq 10.10}$$

where:

- C = the price of a call option
- P = the current price of the underlying asset
- E = the exercise price
- e = the base of natural logarithms
- r = the continuously compounded interest rate
- t = the remaining life of the call option

$N(d_1)$ and $N(d_2)$ are the cumulative probabilities from the normal distribution of getting the values d_1 and d_2, where d_1 and d_2 are as follows:

$$d_1 = \frac{\ln(P/E) + (r + 0.5\sigma^2)t}{\sigma\sqrt{t}}$$

$$d_2 = d_1 - \sigma\sqrt{t}$$

where σ = the standard deviation of the continuously compounded rate of return on the underlying asset.

The term e^{-rt} is the present value of $1 received t periods from the present. It is the continuously compounded equivalent of what we have called d, the present value of $1.

To understand the Black–Scholes model better, consider the case where $N(d_1)$ and $N(d_2)$ are both equal to 1. This is equivalent to assuming complete certainty. Then the model becomes Equation 10.11

$$C = P - Ee^{-rt} \qquad \text{Eq 10.11}$$

$N(d_1)$ and $N(d_2)$ represent cumulative probabilities from the normal distribution. Figure 10.5 illustrates these cumulative probabilities, which must be numbers between 0 and 1. If they

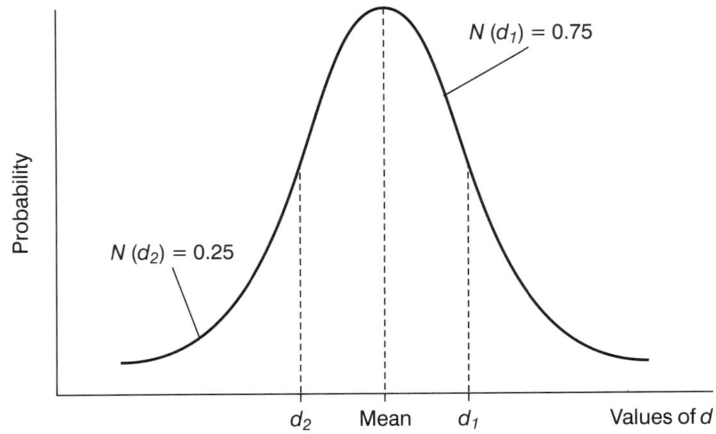

Figure 10.5 Cumulative normal distribution

are less than 1.0, there is some uncertainly about the level of the stock price at option expiration. From the definition of d_1,d_2 must be smaller than d_1. Assume that we know that $N(d_1)$ is 0.75 and $N(d_2)$ is 0.25, then the Black–Scholes model becomes Equation 10.11:

$$C = (0.75)\ P - (0.25)Ee^{-rt} \qquad \text{Eq 10.12}$$

The Black–Scholes function is shown in Figure 10.6.

Since the Black–Scholes model requires no cash payments and interest rate certainty, it cannot be applied to debt instruments. However, the Black–Scholes models can be applied to common stocks without dividends. The model can be adapted for common stocks that pay dividends.

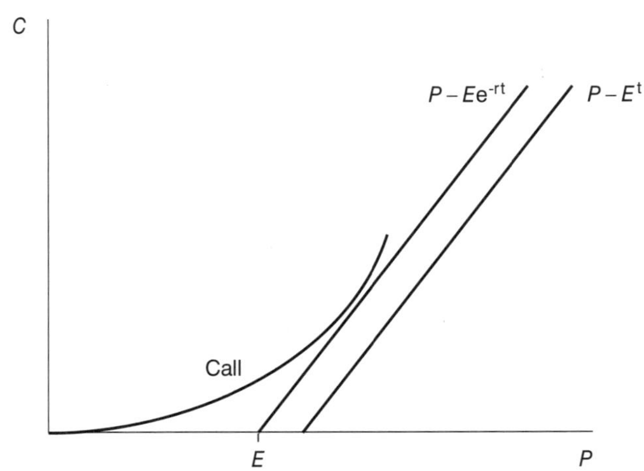

Figure 10.6 Black–Scholes model

Real options and Internet and technology stock valuations

Real options as a method of equity valuation have recently gained much attention. A real option is a choice in the management of real (as opposed to financial) assets or operations, such as the option to accelerate or defer capital expenditure, to change process inputs or outputs, to build or abandon capacity or to alter the nature or to change the scale of a production process. But why do we apply the term real options for investment decisions when traditionally real investment decisions are traditionally analysed using discounted cash flow techniques?

How does an investment project become an option?

A corporate investment opportunity is like a call option because the corporation has the right, but not the obligation, to acquire something, let us say, the operating assets of a new business. If we could find a call option sufficiently similar to the investment opportunity, the value of the option would tell us something about the value of the opportunity. Unfortunately, most business opportunities are unique, so the likelihood of finding a similar option is low. The only reliable way to find a similar option is to construct one.

To do this, we need to establish a correspondence between the project's characteristics and the five variables that determine the value of simple call option on a stock. By mapping the characteristics of the business opportunity onto the template of a call option, we can obtain a model of the project that combines its characteristics with the structure of a call option. Assume the option we will use is a *European* call, which is the simplest of all options because it can be exercised on only one date, its expiration date. The option synthesized here is not a perfect substitute for the real opportunity, but it is still informative. Table 10.9 shows the correspondence illustrating the similarities.

Many projects involve spending money to buy or build a productive asset. Spending money to exploit such a business opportunity is analogous to exercising an option on, for

Table 10.9 Mapping an investment opportunity onto a call option

Investment opportunity	*Variable*		*Call option*
Present value of a project's operating assets to be acquired	S	⇒	Stock price
Expenditure required to acquire the project assets	X	⇒	Exercise price
Length of time the decision may be deferred	t	⇒	Time to expiration
Time value of money	r_f	⇒	Risk-free rate of return
Risk of the project assets	σ^2	⇒	Variance of returns on stock

example, a stock. The amount of money expended corresponds to the option's *exercise price* (denoted for simplicity as *X*). The present value of the asset built or acquired corresponds to the *stock price* (*S*). The length of time the company can defer the investment decision without losing the opportunity corresponds to the option's *time to expiration* (*t*). The uncertainty about the future value of the projects cash flows (that is, the riskiness of the project) corresponds to the *standard deviation of returns* on the stock (*i*). Finally, the time value of money is given in both cases by the *risk-free rate of return* (r_f). By pricing an option using values for these variables generated from our project, we learn more about the value of the project than a simple discounted cash flow analysis can tell us.

Linking NPV and option value

Traditional DCF methods would assess an investment opportunity by computing its net present value (NPV), which is the difference between how much the operating assets are worth (their present value) and how much they cost:

$$\text{NVP} = \text{Present value of assets} - \text{Required capital expenditure}$$

When NPV is positive, the corporation will increase its own value by making the investment. When NPV is negative, the corporation is better off not making the investment. When a final decision on the project can no longer be deferred; that is, when the company's 'option' has reached its expiration date, the project's option value and NPV are the same. At that time, either

$$\text{the option value} = S - X$$

or

$$\text{the option value} = 0$$

whichever is greater. But note that

$$\text{NPV} = S - X$$

as well, because we know from Table 10.9 that S corresponds to the present value of the project assets and X to the required capital expenditure. To reconcile the two completely, we need only observe that when NPV is negative, the corporation will not invest, so the project value is effectively zero (just like the option value) rather than negative. In short, both approaches boil down to the same number and the same decision.

This common ground between the NPV and the option value has great practical significance. It means that corporate spreadsheets set up to compute conventional NPV are highly relevant for option pricing. Any spreadsheet that computes NPV already contains the information necessary to compute S and X, which are two of the five option-pricing variables. Accordingly, executives who want to begin using option pricing need not discard their current DCF-based systems.

NPV and option pricing do diverge, however, when the investment decision may be deferred. The possibility of deferral gives rise to two additional sources of value. First, we would always rather pay for assets later than sooner, all else being equal, because we can earn the time value of money on the deferred expenditure. Second, while we're waiting, the world can change. Specifically, the value of the operating assets we intend to acquire may change. If their value goes up, we haven't missed out; we can still acquire them simply by making the investment (exercising our option). If their value goes down, we might decide not to acquire them. That also is fine (very good, in fact) because, by waiting, we avoid making what would have turned out to be a poor investment. We have preserved the ability to participate in good outcomes and insulated ourselves from some bad ones.

For both of these reasons, being able to defer the investment decision is valuable. Traditional NPV misses the extra value associated with deferral because it assumes the decision cannot be deferred. In contrast, option pricing *presumes* the ability to defer and provides a way to quantify the value of deferring. So to value the investment, we need to develop new metrics that capture these extra sources of value.

Real option pricing methods

In their book *Investment Under Uncertainty*, Dixit and Pindyck (1994) use a simple example to illustrate the difference between NPV and option pricing methods.

Suppose you are deciding whether to invest \$1600 in a new project that makes widgets. The cash flow per widget is \$200 but will change to either \$300 or \$100 at the end of the year with equal probability. After that it will stay at its new level forever. Note that the expected future cash flow is \$200, the weighted average of the risky outcomes, \$300 and \$100. The cost of capital is 10%. Assuming that one widget can be sold immediately, and one per year thereafter, the net present value of the project would be estimated as follows:

$$\text{NPV} = \text{MAX}\left[-1600 + \sum_{t=0}^{\infty}, 0\right] \frac{200}{(1.1)^t}$$

$$= \text{MAX}\left[(-1600 + 2200), 0\right]$$

$$= 600$$

The NPV approach discounts the expected project cash flow at the weighted average cost of capital. The decision rule is to take the maximum of the discounted expected cash flows or zero (meaning don't do the project). The NPV rule is the maximum (determined today) of the expected values. It also makes the implicit assumption that the project should be undertaken immediately or not at all, because the maximum must be decided right now. This assumption rules out the possibility of deferring the investment one year until the uncertainty about the price per widget is resolved. If we look at the project given the option to defer, the economics look better:

$$\text{Option value} = \left[0.5\ \text{MAX}\left(\frac{-1600}{1.1}\sum_{t=1}^{\infty} + \frac{300}{(1.1)^t}\right)\right], 0$$

$$\left[+ 0.5\ \text{MAX}\left(\frac{-1600}{1.1}\sum_{t=1}^{\infty} + \frac{100}{(1.1)^t}\right)\right], 0$$

$$= 0.5\ \text{MAX}\left(\frac{-1600 + 3300}{1.1}, 0\right)$$

$$+ 0.5\ \text{MAX}\left(\frac{-1600 + 1100}{1.1}, 0\right)$$

$$= 0.5\left(\frac{1700}{1.1}\right) + 0.5(0)$$

$$= 7.33$$

With the option to defer you can wait one period to invest, then decide whether to do so, based on the arrival of information about the long-term cash flow per widget. If the cash flow is only 100 per unit you will not exercise the option to invest, but if the cash flow is 300 per unit, you will. Although the NPV of investing immediately is \$600, the NPV should you decide to defer is even higher at \$733. Therefore you will defer.

The value of this call option (with an exercise price of \$1600, a one-year life, a variance determined by the cash-flow spread of \$200 per unit, and an underlying risky asset that has a value without flexibility of \$600) is the difference between the value of the project with flexibility and its value without flexibility, \$733 – \$600 = \$133. Note also that the NPV is the maximum, decided today, of the expected discounted cash flows or zero, while the option value is the expected value of the maximums, decided when information arrives, of the discounted cash flows in each future state of nature, or zero:

$$\text{NPV} = \underset{t=0}{\text{MAX}}\left(\frac{\text{Expected cash flows}}{\text{Cost of capital}}\right), 0$$

$$\text{Option value} = \text{Expected}\left[\underset{t=t}{\text{MAX}}\left(\frac{\text{Cash flow given info}}{\text{Cost of capital}}\right), 0\right]$$

The two methods use information quite differently. NPV forces a decision based on today's expectation of future information, while option valuation allows the flexibility of making decisions in the future contingent on the arrival of information. Option-pricing methods capture the value of flexibility while NPV does not. So the real options approach formalizes and captures in a mathematical sense what is already reflected in valuations in a more subjective way. Analysts and investors recognize that flexibility – the ability to react to changing circumstances and new information – is linked to value. That premium is effectively the option value.

Types of real options

There are many types of real options and methods of classifying them. The categories below are designed to capture the key sources of option value in businesses. Three real options are discussed below: scale, scope and timing options.

Scale (growth) options

Scale or growth options represent the value inherent in future investment opportunities that expand existing business operations. Once a company is committed to a new investment then the growth option value disappears (although this may be replaced by other options such as the option to abandon or curtail operations) and the value of the investment opportunity is simply the expected present value of the investment cash flows.

Growth *options* need to be differentiated from growth *opportunities.* Many growth opportunities do not have any option value since there may be significant upfront commitment and little in the way of learning benefits. In this case, the value of the growth opportunity is simply the expected NPV. Growth options are more valuable since they give a company the opportunity to use flexibility in their operations to benefit from uncertainty.

Scope options

These are the options available to companies that enter new markets, develop new business models and reinvent themselves. Although growth options are difficult to identify and value, scope options are even more difficult. Clearly there is a value inherent in the possibility that a company with an existing position in a market, a valuable customer base, and so on, will be in a better position than others to take advantage of new business opportunities. These opportunities are most prevalent in a sector where there is rapid development through technological change – the Internet being an obvious illustration of these conditions.

In addition to the effects of technological change, the reason why scope options are particularly important for Internet stocks is that their value is much more dependent upon information. Established Internet companies derive vast quantities of data from the on-line activities of their customers and this, when managed effectively, should reduce the risks inherent in new business activities.

Timing options

Flexibility, and the ability to react appropriately as a company learns from new information acquired, is the key to realizing real option value. A timing option arises where a company can use information to time the entry into a new venture (or the exit from an activity) to best effect.

The link between real options and DCF

It is very important to remember that the assumptions and inputs in a real options valuation include all those necessary for a discounted cash flow valuation plus several others. Real options methodology is inevitably more difficult to apply in the real world than DCF.

In many cases, although investment opportunities have intrinsic (or more correctly forward) value, they nevertheless have little or no option value. The time value of the option is zero. There are three main circumstances where this applies.

(1) The investment opportunity is significantly 'in the money'

If an investment opportunity is expected to be highly profitable and the chance of it not being so profitable (and of the company being able to curtail the investment to avoid loss) is remote, then the options value is likely to be material. This effect is easily seen in the equity options market where call options with a strike price significantly below the share price essentially trade at their intrinsic value (their NPV).

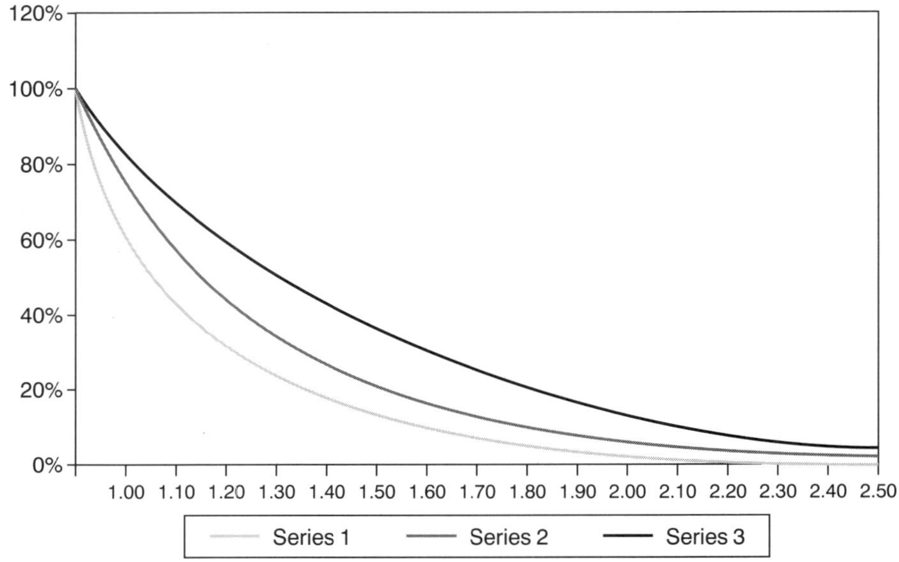

Figure 10.7 Contribution of real options to investment valuation (real option as a percentage of total investment value). Note: The chart is derived from a simple Black–Scholes option valuation assuming a decision period of 2 years during which there is perfect flexibility. The three lines represent volatility of 20%, 40% and 60% respectively. *Source:* UBS Warburg

Figure 10.7 illustrates this by showing the theoretical value (see other limitations below) of a real option assuming a certain profitability index for an investment opportunity. A profitability index is the ratio of the gross present value of expected cash flows of an investment opportunity to the investment required.

An alternative way of presenting this data is to show the absolute net present value data split into the expected intrinsic value of the investment and the time value of the option (see Figure 10.8).

Figures 10.7 and 10.8 show that, even where there is high volatility, the option value only has a limited contribution to overall investment value if the investment is expected to be highly profitable (in this analysis a high-profitability index). However, real options can contribute significantly to value where investment decisions are more marginal.

(2) There is limited flexibility to react to new information

Options have time value because investors have the opportunity to modify their actions to their best advantage. In a traded

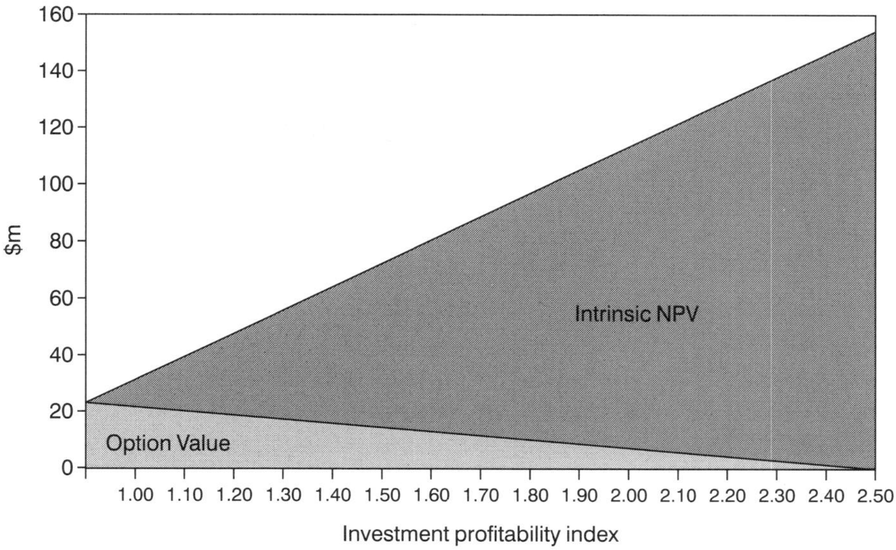

Figure 10.8 Analysis of investment NPV. Note: This chart uses a 40% volatility measure but otherwise is based upon the same data as the previous chart. *Source:* UBS Warburg

equity option, the investor can decide whether or not to invest (to exercise the option or not) up until the final exercise date, at which point it can be observed, with certainty, whether exercise is profitable. Action can be modified perfectly always to take best advantage of the circumstances faced. Now imagine the option holder has to commit to exercising or not when the option is first acquired. Of course it is then not an option at all; it is simply a future and would attract an option premium. Valuation would be at the present value of the value of the investment. Option value critically depends upon the ability to modify strategy, i.e. change from exercise to not exercise as appropriate depending on changes in circumstances (in the case of the traded equity option, the movements in the stock price).

(3) Management does not have the ability to react appropriately

Although there may be the potential of real option value due to the opportunities available to a company and the flexibility to react to these, the actual realization of this value depends on the ability of management. Only if management is capable of recognizing the existence of options and only if they are capable of effectively using information to exercise the options appropriately is a real option premium justified. This is the reason that investors are willing to pay more for companies with 'good management' in place.

How does this apply to real corporate investment opportunities? For a future business opportunity to have option value (i.e. value in excess of the intrinsic value – the expected NPV), there must be the opportunity for the company to modify its strategy to the best effect.

11

What are the lessons for investors from the year 2000 technology stocks collapse?

The concept of the New Economy, or the 'new economic paradigm' as it was sometimes referred to, contained a number of distinct strands, some true, some half true, and some simply wrong. All these ideas came together in the US economy of the second half of the 1990s, as we discussed in Chapter 2. To distinguish what may still be true from what we now know to be false, that skein of ideas must be unravelled.

On analysis, four narrow notions emerged; that the business cycle was dead; that capital markets now worked in unprecedented ways; that we were living in an era of unique technical progress; and, finally, that information technology had fundamentally altered the way the economy works.

The business cycle is decidedly alive. That should not surprise anybody. Yet the view that business cycles were dead rested on several correct presumptions about the US economy: labour markets had become more competitive; global competition had reduced the pricing power of both trade unions and big business; the need for inventories had diminished; and monetary policy was far more credibly oriented towards price stability than in the inflationary 1960s and 1970s.

What this analysis overlooked, however, is that cycles can occur and frequently have occurred in flexible economies with credible monetary anchors. All that is needed is a big investment expansion fuelled by expanding credit and a strong stock market. Indeed, the firmer the belief that the business cycle is dead, the greater the confidence and the likelihood of business cycles.

The capital markets have certainly hummed to a different tune in the 1990s. But Internet mania looks like an only slightly less irrational version of the South Sea Bubble. Some rationalizations for the prices of stocks were as ludicrous as the prices themselves. Nor was bubble analysis restricted to the dot.coms. Just look at telecommunications.

Yet behind this ludicrous hype, important changes did take place. In particular, the rise of venture capital had made it far easier for entrepreneurs to create valuable new companies. While capital markets are as prone as ever to excessive swings between optimism and pessimism, they do also seem better able to support and promote innovation.

The notion that information technology represents the greatest transformation since the industrial revolution is historically illiterate. The technological changes between 1880 and 1940 exceeded, in both scope and intensity, all that has happened since. Those changes included new sources of energy (electricity and petroleum), new industries (motor vehicles and pharmaceuticals) and new products (cars, washing machines, telephones, radio, television, penicillin). These profoundly altered what was produced and how. They also transformed the way people lived. Against all that, what are the personal computer and even the Internet?

Even if information technology is not more important than all the innovations that went before it, as some believe, the rise of the 'weightless economy' does have revolutionary implications. It makes information and knowledge the most important economic commodity.

But information, as economists have long known and many businesses are finding out for themselves, is a peculiar commodity. If the marginal cost of providing an additional customer with a commodity is zero, standard models of a competitive economy start to break down. This change creates huge challenges for traditional business models, for the organization of companies and for public policy in relation to intellectual property and competition.

The laws of economics have not been repealed

It has finally being recognized that all technology companies are subject to the same laws of economics as other businesses. These laws, which set strict long-term limits to profitability and therefore sustainable stock market valuations, ultimately brought all technology stocks back down to earth.

Specifically, there are three economic principles that stock market analysts continuously overlooked in the tech stock mania.

The first is simply the law of supply and demand. In any competitive economy new capital will flow into any industry that enjoys very high profits and growth. This flow will continue until the risk-adjusted return on capital falls to the economy's normal or average rate. Super-normal profits can be maintained only over long periods in markets that are not competitive, where individual companies enjoy a large degree of monopoly power. But experience shows that there have been few monopolies in technology industries.

The second economic principle that Silicon Valley is now recognizing, and investors ignored at their peril, is the law of diminishing returns. Technology may continue to advance as rapidly as ever – for example, the speed of microprocessors is still doubling every 18 months – but every new leap in technology is likely to be less economically valuable than the one before. Once a computer processor reaches a certain power further additions to its speed or its storage capacity will not offer much economic value to users. The same is likely to prove true of broadband Internet connections and third generation mobile phone services. These will doubtless improve services, but will consumers pay large premiums for these improvements to services they already regard as satisfactory?

Many investors believed that the Internet had not only repealed, but also actually reversed, the law of diminishing returns. The Internet was supposed to be subject to a new 'network effect', which asserted that each new addition to the network, every new subscriber and every new service, would generate more profit than the one before. This was obviously true up to a point, since the Internet, like any other business, required a certain critical size to start enjoying the economies of mass merchandising and mass production. But once the Internet passed the critical point where it became a genuine mass global network, additional subscribers have become less wealthy, more price sensitive and less commercially valuable. In short, the law of diminishing returns has come back into force.

The third economic law that investors have started to rediscover is the time value of money. Profits in the future are worth far less than cash in the hand today. Technology investors habitually talk about the amazing speed of technical and commercial change in the Internet world. Yet the way they valued technology stocks assumed that revenues and profits would continue to grow without interruption for 20 or more years!

Recent mistakes must be avoided

As mentioned above there is no doubt that the computer revolution has infiltrated almost every part of the economy, greatly increasing potential efficiency. In the US the productivity of labour is probably capable of rising by 2.5% a year, compared with 1.5% in recent decades.

There is good reason to believe that this improvement can be sustained. The doubling in complexity of computer chips every 18 months will continue for some time. Communications costs will fall further. As happened after the development of electric motors, human ingenuity will find new ways to refine and apply these technologies.

But investors and bankers must avoid two big mistakes of recent years. First, they must not expect high technology to translate automatically into high demand. The standard investment in mobile phones shows that an extra ingredient is needed: the tailoring of technology to real human needs – and they are notoriously unpredictable.

Second, investors must be under no illusion that high technology will improve general profit levels. It will not. The advantage of improved efficiency will be eroded by competition. And the race to the winning-post in high-tech investments will leave many competitors floundering in the mud.

Schumpeter and creative destruction

All attempts to understand the effects of technological progress on economic growth pay homage to Joseph Schumpeter, the Austrian economist discussed in Chapter 2, best remembered for his views on the 'creative destruction' associated with industrial cycles 50–60 years long. Arguably the most radical economist of the twentieth century, Schumpeter was the first to challenge classical economics as it sought (and still seeks) to optimize existing resources within a stable environment – treating any disruption as an external force on a par with plagues, politics and the weather. Into this intellectual drawing room, Schumpeter introduced the raucous entrepreneur and his rumbustious behaviour. As Schumpeter saw it, a normal healthy economy was not one in equilibrium, but one that was constantly being 'disrupted' by technological innovation.

But will the market forget these simple laws of economics? It is wise to remember the words of Alan Greenspan, Chairman of the Federal Reserve, speaking in early 2000:

> History tells us that sharp reversals in confidence happen abruptly, most often with little advance notice . . . Claims on far distant future values are discounted to insignificance. What is so intriguing is that this type of behaviour has characterized human interaction with little appreciable difference over the generations. Whether Dutch tulip bulbs or Russian equities, the market price patterns remain much the same.

Bibliography

Angel, J.A. @www.msb.edu/faculty. Georgetown University, Net Stock Valuation Calculator.

Aswath, D. (1996) *Investment Valuation: Tools and Techniques for Determining the Value of an Asset.* John Wiley & Sons.

Berners-Lee, T. (1999) *Weaving the Web.* HarperCollins.

Bernstein, P.L. (1998) 'Are Networks Driving the New Economy?' *Harvard Business Review*, November–December, pp. 159–66.

Black, F. and Scholes, M.S. (1972) 'The Valuation of Options Contracts and a Test of Market Efficiency', *Journal of Finance*, 27: 399–417.

Black, F. and Scholes, M.S. (1973) 'The Pricing of Options and Corporate Liabilities', *Journal of Political Economy*, 81: 637–54.

Blanchard, Olivier (1990) 'Why Does Money Affect Output?'. In Benjamin Friedman and Frank Hahn (eds), *Handbook of Monetary Economics*. Elsevier, pp. 780–835.

Blanchard, Olivier (1993) 'Movements in the Equity Premium', *Brooking Papers on Economic Activity*, 2: 75–138.

Blanchard, Olivier, J. and Watson, M.W. (1982) 'Bubbles, Rational Expectations and Financial Markets'. In P. Watchel (ed), *Crisis in the Economic and Financial System.* Lexington Books.

Blinder, Alan (1997) 'Is There a Core of Practical Macroeconomics that We All Should Believe?', *American Economic Review*, 87: 240–3.

Blinder, A. (2000) 'The Internet and the New Economy'. Progressive Policy Institute, January, www.ppionline.org

Business 2.0 (2000) The 10 Essential Net Business Principles. June 2000.

Cerf, V. and Navasky, J. *A Brief History of the Internet and Related Networks* gopher://gopher.isoc.org:70/00/internet/history

Cochrane, J. (1997a) 'Where Is the Market Going? Uncertain Facts and Novel Theories', *Economic Perspectives*, Federal Reserve Bank of Chicago, 21 (6):3–37.

Cochrane, J. (1997b) 'New Facts in Finance,' *Economic Perspectives,* Federal Reserve Bank of Chicago, 21 (6): 36–58.

Copeland, T., Koller T. and Murrin, J. (2000) *Valuation, Measuring and Managing the Value of Companies*, 3rd edn. John Wiley & Sons.

Coyle, D. (1998) *The Weightless World: Strategies for Managing the Digital Economy.* MIT Press.

Cuthbertson, K. (1998) *Quantitative Financial Economics, Stocks, Bond, and Foreign Exchange.* John Wiley & Sons.

Dixit, A.K. and Pindyck, R.S. (1994) *Investment under Uncertainty.* Princeton University Press.

Estes, Ralph (1981) *Dictionary of Accounting.* MIT Press, pp. 81, 105.

Flood, Robert P. and Garber, P.M. (eds) (1994) *Speculative Bubbles, Speculative Attacks, and Policy Switching.* MIT Press.

Galbraith, S.K. (1955) *The Great Crash 1929.* Houghton Miflin.

Garber, Peter M. (1989a) 'Tulipmania', *Journal of Political Economy*, 97 (31): 535–60.

Garber, P.M. (1989b) 'Who Put the Mania in Tulipmania?', *Journal of Portfolio Management*, 16 (1): 53–60.

Garber, P.M. (1990) 'Famous First Bubbles', *Journal of Economic Perspectives*, 4 (2).

Garber, P.M. (2000) *Famous First Bubbles: The Fundamentals of Early Manias.* MIT Press.

Gordon, R.J. (1997) 'The Time Varying NAIRU and its Implications for Economic Policy', *Journal of Economic Perspectives*, 11 (1): 11–32.

Gordon, R.J. (1998a) 'Foundations of the Goldilocks Economy: Supply Shocks and the Time-Varying NAIRU', Brookings Papers on Economic Activity, no.2, pp. 297–346.

Gordon, R.J. (1998b) 'Monetary Policy in the Age of Information and Technology: Computers and the Solow Paradox', manuscript.

Gordon, R.J. (1999) 'Has the "New Economy" Rendered the Productivity Slowdown Obsolete', Working Paper, Northwestern University.

Graham, B. (1949) *The Intelligent Investor.* Harper & Brothers (1985 edn, HarperCollins).

Graham, B. and Dodd, D. (1934) *Securities Analysis.* McGraw Hill

Greenspan, A. (1997) 'Monetary Policy Testimony and Report to Congress'. Subcommittee on Domestic and International Monetary Policy on Banking and Financial Services, US House of Representatives, 22 July 1997.

Greenspan, A. (1998) 'Question: Is There a New Economy?', remarks at the University of California, Berkeley, 4 September 1998.

Greenspan, A. (2000) 'The Evolving Demand for Skills', at the US Department of Labor National Skills Summit, Washington, DC, 11 April 2000. http://www.federalreserve.gov/boarddocs/speeches/2000/20000411.htm

Greenspan, A. (2000) 'Retail Payment Systems', before the National Automated Clearinghouse Association Annual Meeting, Los Angles, California (via videoconference), 10 April 2000. http://www.federalreserve.gov/boarddocs/speeches/2000/20000410.htm

Greenspan, A. (2000) 'Technology Innovation and Its Economic Impact', before the National Technology Forum, St Louis, Missouri (via videoconference), 7 April 2000. http://www.bog.frb.fed.us/boarddocs/speeches/2000/20000407.htm

Greenspan, A. (2000) 'Technological Innovation and the Economy', before the White House Conference on the New Economy, Washington, DC, 5 April 2000. http://www.bog.frb.fed.us/boarddocs/speeches/2000/20000405.htm

Greenspan, A. (2000) 'Evolution of Our Equity Markets', before the Committee on Banking, Housing, and Urban Affairs, US Senate, 13 April 2000. http://www.federalreserve.gov/boarddocs/testimony/2000/2000413/htm

Hardouvelis, G.A. (1988) 'Evidence on Stock Market Speculative Bubbles: Japan, United States and Great Britain', Federal Reserve Bank of New York Research Paper No. 8810.

Harmon, Steve (1998) 'Home Improvement Web Style: Inside the Internet Valuation Tool Chest', Morning Report, Internet Stock Report, 14 July 1998. www.internetstockreport.com/column/article/0,1785,57651,00.html

Herzenberg, S.A., Alice, J.A. and Wial, H. (1999) *New Rules for a New Economy: Employment and Opportunity in Post-Industrial America.* Cornell.

Jones, C.P. (1999) *Investment, Analysis and Management.* John Wiley & Sons.

Jorgenson, D. and Stiroh, K. (1999) 'Information Technology and Growth', *American Economic Review, Papers and Proceedings*, 89: 109–22.

Jorgenson, D. and Stiroh, K. (2001) 'What Drives Productivity Growth?', *Federal Reserve Bank of New York Policy Review*, June 2001.

Kahn, R.E. and Cerf, V.G. (1999) 'What is the Internet (And What Makes It Work)?', paper prepared for the Internet Policy Institute.

Kelly, K. (1998) *New Rules for the New Economy: 10 Radical Strategies for a Connected World*. Viking Press.

Kettell, B. (1998) *What Drives Financial Markets?* Financial Times/Prentice Hall.

Kettell, B. (1999) *Fed Watching*. Financial Times/Prentice Hall.

Kettell, B. (2000) *What Drives Currency Markets?* Financial Times/Prentice Hall.

Kettell, B. (2001) *Financial Economics*. Financial Times/Prentice Hall.

Keynes, J.M. (1936) *The General Theory of Employment, Interest and Money*. Harcourt.

Kindleberger, C. (1978) *Manias, Panics, and Crashes: A History of Financial Crises*. Basic Books.

Krugman, P. (1997) 'How Fast Can the US Economy Grow?', *Harvard Business Review*, 75 123–9.

Law, John (1934 [1705]) 'Money and Trade Considered with a Proposal for Supplying the Nation with Money' (P. Horsin, ed.), *Oeuvres Complètes*, vol. 1. Paris.

Luehrmann, T.A. (1998) 'Investment Opportunity as Real Options: Getting Started on the Numbers', *Harvard Business Review*, July–August.

Livingston, M. (1996) *Money and Capital Markets*, 3rd edn. Blackwell Business.

Lynch, R. (1997) *Corporate Strategy*. Pitman.

Mackay, C. (1841) *Memoirs of Extraordinary Popular Delusions and the Madness of Crowds*. Bentley.

Malkiel, B.G. (2000) *A Random Walk down Wall Street*, 10th edn. W.W. Norton.

McTeer, R.D. (1999) 'The New Paradigm', Annual Report of the Federal Reserve Bank of Dallas.

Meyer, L.H. (1997) 'Statement on Monetary Policy', Committee on Banking and Financial Services, US House of Representatives, 23 July.

Meyer, L.M. (2000) 'The Economic Outlook and the Challenges Facing Monetary Policy' given before the Toronto

Association for Business and Economics, Toronto, Canada, 12 April 2000. http://www.federalreserve.gov/boarddocs/speeches/2000/20000412.htm

Modigliani, F. and Papademos, L. (1975) 'Targets for Monetary Policy in the Coming Year', Brookings Papers on Economic Activity, 1, pp. 141–63.

Nilson, T. (1998) *Competitive Branding.* John Wiley & Sons.

Nakamura, L. (1999) 'Intangibles: What Put the New in the New Economy?' *Federal Reserve Bank of Philadelphia Business Review*, July/August.

O'Shaughnessy, James P. (1996) *What Works on Wall Street.* McGraw-Hill.

Perkins, M. and Perkins, A. (1999) *The Internet Bubble: Inside the Overvalued World of High-Tech Stocks and What You Need To Know To Avoid the Coming Shakeout.* Harper Business.

Phillips, A.W. (1958) 'The Relation Between Unemployment and the Rate of Change of Money Wage Rates in the United Kingdom, 1861–1957', *Economica*, November, pp. 283–99.

Porter, M.E. (1985) *Competitive Advantage: Creating and Sustaining Superior Performance.* The Free Press.

Porter, M.E. (1998) *Competitive Strategy: Techniques for Analyzing Industries and Competitors.* The Free Press.

Schiller, R.J. (1986) Survey Evidence Regarding the September 11–12 Stock Market Drop. Yale University.

Schiller, R.J. (1987) Survey Evidence Regarding Investor Behavior in the October 1987 Stock Market Crash. Yale University.

Schiller, R.J. (2000) *Irrational Exuberance.* Princeton University Press.

Schiller, R.J. (2000) 'Measuring Bubble Expectations and Investor Confidence', *Journal of Psychology and Markets*, 1 (1): 49–60.

Schiller, R. and Siegel, J. (1987) 'The Gibson Paradox and Historical Movements in Rail Interest Rates', *Journal of Political Economy*, 89 (5).

Schumpeter, J. (1927) 'The Explanation of the Business Cycle', *Economica*, 7 (21): 295.

Schumpeter, J. (1934) *The Theory of Economic Development.* Harvard University Press.

Schumpeter, J. (1939) *Business Cycles.* McGraw Hill.

Schumpeter, J. (1951) 'The Creative Response in Economic History'. In Richard Clemence (ed.), *Essays on Economic Topics of J.A. Schumpeter.* Kennikat Press.

Schwartz, E.S. and Moon, Mark (2000) *Rational Pricing of Internet Stocks*. Association for Investment Management and Research.

Sichel, Daniel E. (1997) *The Computer Revolution*. Brookings.

Siegel, J. (1995) *Stocks for the Long Run: A Guide to Selecting Markets for Long-Term Growth*. Irwin Professional Publishing.

Solow, R. (1956) 'A Contribution to the Theory of Economic Growth', *Quarterly Journal of Economics*, 70: 65–94.

Smithers, A. and Wright, S. (2000) *Valuing Wall Street: Protecting Wealth in Turbulent Markets*. McGraw Hill.

Sperling, J.G. (1962) *The South Sea Company*. Baker Library.

Staiger, D., Stock, J.H. and Watson, M.W. (1997) 'The NAIRU, Unemployment and Monetary Policy', *Journal of Economic Perspectives*, 11(1): 33–50.

Stock, J.H. (1998) 'Monetary Policy in a Changing Economy: Indicators, Rules, and the Shift Towards Intangible Output', Harvard University, October.

Timmets, J. (1999) *Electronic Commerce*. John Wiley.

Toffler, A. (1980) *The Third Wave*. Morrow.

UBS Warburg Global Equity Research (May 2000) 'Navigating the I-Valuation Jungle'.

Wadhwani, S. (2000) 'Monetary challenges in a "New Economy"', *Bank of England Quarterly Bulletin*, 40 (4): 411–22.

Williams, J.R. (1999) *Miller GAAP Guide*. Harcourt Brace.

Websites

www.ft.com
www.nasdaq.com
www.economist.com
www.Internet.com
www.thestreet.com
www.fiepr.com
www.businessweek.com
www.businessage.com
www.financial-planner.com
www.Kiplinger.com
www.Money.com
www.infoworld.com
www.cybervaluation.com
www.bbc/new.co.uk
www.bloomberg.com
www.techweb.com
www.wileyvaluation.com
www.expressindia.com
www.redherring.com
www.moneycentral.msn.com
www.itulip.com
www.netratings.com
www.stocks.about.com
www.nielsen-netratings.com
www.forbes.com
www.techweek.com
www.swoba.hhs.se
www.ic24.moneyworld.co.uk
www.online-stock-market-quotes.com
www.internetstockreport.com
www.internet.idg.net
www.morningstar.com
www.bigcharts.com
www.europe.cnnfn.com
www.multexinvestor.com
www.uk.mmxieurope.com
www.stockhouse.com
www.stocksontheweb.com
www.forrester.com
www.stern.nyu.edu
www.ubswarburg.com

www.valuengine.com
www.investorwords.com
www.adventas.com
www.morganstanley.com
www.babson.edu
www.fortressdesing.com
www.jpmorgan.com
www.wdr.com
www.yahoo/finance.com
www.siliconinvestor.com
www.dallasfed.com
www.FRB.com
www.Nasdaq.com
www.yardeni.com
www.dowjones.com
www.wsj.com
www.csfb.com
www.tprice.com
www.investorguide.com
www.internationalreport.com
www.worth.com
www.yahoo.com
www.amazon.com
www.amcity.com
www.netconductor
www.zdnet.com

Index